AF886492

U.S. Marine Corps Close Combat

Plus

Marine Pistol Marksmanship

Department of Defense

Pentagon
Publishing

DEPARTMENT OF THE NAVY
Headquarters United States Marine Corps
Washington, D.C. 20380-1775

18 February 1999

FOREWORD

1. PURPOSE

Today's Marines operate within a continuum of force where conflict may change from low intensity to high intensity over a matter of hours. Marines are also engaged in many military operations other than war, such as peacekeeping missions or noncombatant evacuation operations, where deadly force may not be authorized. During noncombative engagements, Marines must determine if a situation warrants applying deadly force. Sometimes Marines must decide in a matter of seconds because their lives or the lives of others depend on their actions. To make the right decision, Marines must understand both the lethal and nonlethal close combat techniques needed to handle the situation responsibly without escalating the violence unnecessarily. Marine Corps Reference Publication (MCRP) 3-02B, *Close Combat,* provides the tactics, techniques, and procedures of Marine Corps close combat. It also provides the doctrinal basis for the Marine Corps Close Combat Training Program (MCCCTP).

2. SCOPE

This publication guides individual Marines, unit leaders, and close combat instructors in the proper tactics, techniques, and procedures for close combat. MCRP 3-02B is not intended to replace supervision by appropriate unit leaders and close combat instruction by qualified instructors. Its role is to ensure standardization and execution of tactics, techniques, and procedures throughout the Marine Corps.

3. SUPERSESSION

MCRP 3-02B supersedes Fleet Marine Force Manual (FMFM) 0-7, *Close Combat,* dated 9 July 1993. There are significant differences between the two publications. MCRP 3-02B should be reviewed in its entirety.

4. WARNING

Techniques described in this publication can cause serious injury or death. Practical application in the training of these techniques will be conducted in strict accordance with approved Entry Level Close Combat, Close Combat Instructor (CCI), and Close Combat Instructor Trainer (CCIT) lesson plans. Where serious danger exists, the reac er is alerted by the following:

WARNING

5. CERTIFICATION

Reviewed and approved this date.

BY DIRECTION OF THE COMMANDANT OF THE MARINE CORPS

J. E. RHODES
Lieutenant General, U.S. Marine Corps
Commanding General
Marine Corps Combat Development Command

DISTRIBUTION: 14400006600

OVERVIEW OF CLOSE COMBAT

1. Purpose of Close Combat

Close combat is the physical confrontation between two or more opponents. It involves armed and unarmed and lethal and nonlethal fighting techniques that range from enforced compliance to deadly force. The purpose of close combat is to execute armed and unarmed techniques to produce both lethal and nonlethal results. Unarmed techniques include hand-to-hand combat and defense against hand-held weapons. Armed techniques include techniques applied with a rifle, bayonet, knife, baton, or any weapon of opportunity.

2. Continuum of Force

Marines will find themselves in both combative and noncombative situations. The threat level in these situations can rise and fall several times based on the actions of both Marines and the people involved. The escalation of force stops when one person complies with the demands imposed by another person. This range of actions is known as a continuum of force. Continuum of force is the concept that there is a wide range of possible actions, ranging from voice commands to application of deadly force, that may be used to gain and maintain control of a potentially dangerous situation (MCO 5500.6_, *Arming of Security and Law Enforcement [EE] Personnel and the Use of Force).* Continuum of force consists of five levels that correspond to the behavior of the people involved and the actions Marines use to handle the situation (see the table below). Close combat techniques are executed in levels three, four, and five.

Level One: Compliant(Cooperative)

The subject complies with verbal commands. Close combat techniques do not apply.

Level Two: Resistant (Passive)

The subject resists verbal commands but complies immediately to any contact controls. Close combat techniques do not apply.

Continuum of Force		
Level	**Description**	**Actions**
1	Compliant (Cooperative)	Verbal commands
2	Resistant (Passive)	Contact controls
3	Resistant (Active)	Compliance techniques
4	Assaultive (Bodily Harm)	Defensive tactics
5	Assaultive (Serious Bodily Harm/ Death)	Deadly force
Note: Shading indicates levels in which Marines use close combat techniques.		

Level Three: Resistant (Active)

The subject initially demonstrates physical resistance. Marines use compliance techniques to control the situation. Level three incorporates close combat techniques to physically force a subject to comply. Techniques include—

- Come-along holds.
- Soft-handed stunning blows.
- Pain compliance through joint manipulation and the use of pressure points.

Level Four: Assaultive (Bodily Harm)

The subject may physically attack Marines, but he does not use a weapon. Marines use defensive tactics to neutralize the threat. Defensive tactics include the following close combat techniques:

- Blocks.
- Strikes.
- Kicks.
- Enhanced pain compliance procedures.
- Nightstick blocks and blows.

Level Five: Assaultive (Serious Bodily Harm/Death)

The subject usually has a weapon and will either kill or seriously injure someone if he is not stopped immediately and brought under control. Typically, to control the subject, Marines apply deadly force through the use of a firearm, but they may also use armed and unarmed close combat techniques.

3. Marine Corps Tactical Concepts

Close combat techniques support the following key Marine Corps tactical concepts. The concepts are not standalone ideas but are to be combined to achieve an effect that is greater than their separate sum.

Achieving a Decision

Achieving a decision is important in close combat. An indecisive fight wastes energy and possibly Marines' lives. Whether the intent is to control an opponent through restraint or defend themselves in war, Marines must have a clear purpose before engaging in close combat and act decisively once engaged.

Gaining an Advantage

A basic principle of martial arts is to use the opponent' s strength and momentum against him to gain more leverage than one's own muscles alone can generate, thereby gaining an advantage. In close combat, Marines must exploit every advantage over an opponent to ensure a successful outcome. This can include employing various weapons and close combat techniques that will present a dilemma to an opponent. Achieving surprise can also greatly increase leverage. Marines try to achieve surprise through deception, stealth, and ambiguity.

Speed

Marines use speed to gain the initiative and advantage over the enemy. In close combat, the speed and violence of the attack against an opponent provides Marines with a distinct advantage. Marines must know and understand the basics of close combat so they can act instinctively with speed to execute close combat techniques.

Adapting

Close combat can be characterized by friction, uncertainty, disorder, and rapid change. Each situation is a unique combination of shifting factors that cannot be controlled with precision or certainty. For example, a crowd control mission may call for Marines to employ various techniques ranging from nonlethal restraint to more forceful applications. Marines who adapt quickly will have a significant advantage.

Exploiting Success

Typically, an enemy will not normally surrender simply because he was placed at a disadvantage. Marines cannot be satisfied with gaining an advantage in a close combat situation. They must exploit any advantage aggressively and ruthlessly until an opportunity arises to completely dominate the opponent. Marines must exploit success by using every advantage that can be gained.

(reverse blank)

Close Combat

Table of Contents

Page

Chapter 5. Throws

Chapter 6. Chokes and Holds

Chapter 7. Ground Fighting

Chapter 8. Nonlethal Techniques

Appendices

CHAPTER 1

FUNDAMENTALS OF CLOSE COMBAT

This chapter describes all techniques for a right-handed person. However, all techniques can be executed from eittwr side.

The Marine is depicted in camouflage utilities. The opponent is depicted without camouflage.

The fundamentals of close combat include ranges, weapons of the body, target areas of the body, and pressure points of the body. These fundamentals form the basis for all close combat techniques. They provide Marines with a common framework regardless of the type of confrontation or the techniques used. If Marines apply these fundamentals properly in a close combat situation, they may save their lives or the lives of fellow Marines.

1. Ranges of Close Combat

Close combat engagements occur within three ranges: long range, midrange, and close range. During any engagement, these ranges may blur together or may rapidly transition from one to another until either the opponent is defeated or the conflict is resolved.

Long Range

During long range engagements, combatants engage each other with rifles, bayonets, sticks, or entrenching tools. See figure below.

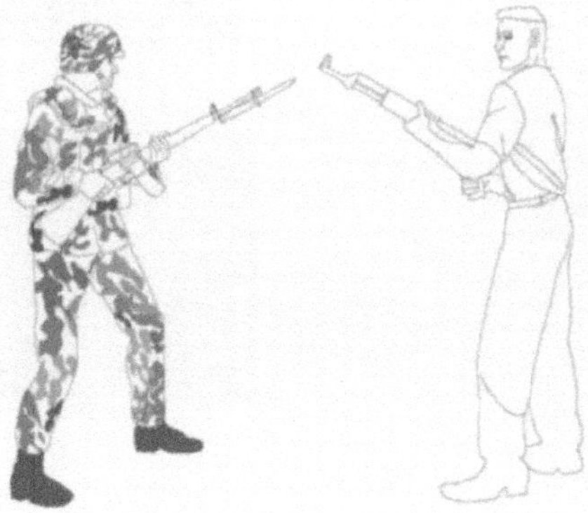

Midrange

During midrange engagements, combatants engage each other with knives, punches, or kicks.

Close Range

During close range engagements, combatants grab each other. Close range engagements also involve elbow strikes, knee strikes, and grappling.

2. Weapons of the Body

Hands and Arms

The hands, forearms, and elbows are the arm's individual weapons. The hands consist of several areas that can be used as weapons: fists, edges of hands, palms, and fingers.

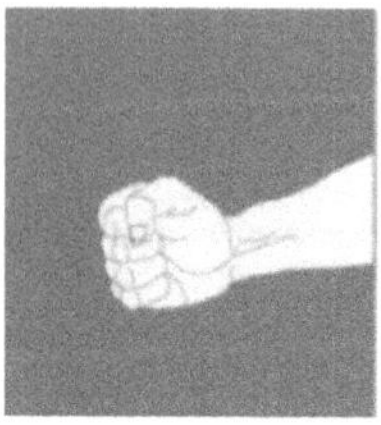

Fists. To minimize injury to the fists, Marines use their fists as weapons to target soft tissue areas such as the throat. The fists' striking surfaces are the first two knuckles of the hands or the meaty portions of the hands below the little fingers.

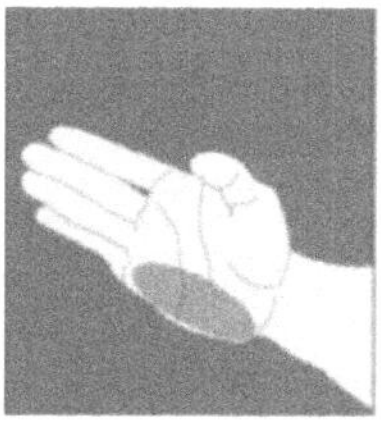

Edge of Hand. Marines use the edge of the hand (knife edge) as a weapon. Marines use the edge of the hand to strike soft tissue areas.

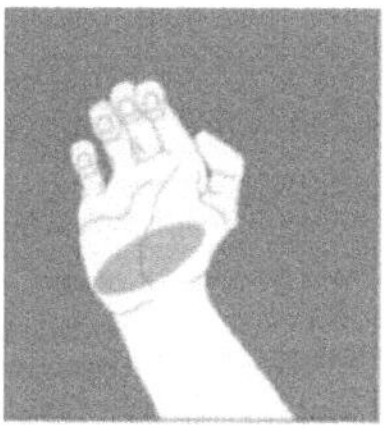

Palms. Because of the palm's padding, Marines use the heels of the palms to strike, parry, and/or block.

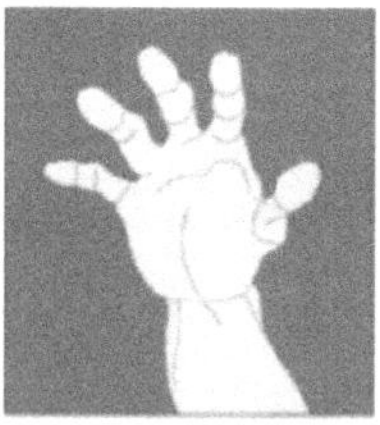

Fingers. Marines use the fingers to gouge, rip, and tear soft tissue areas (e.g., eyes, throat, groin).

Forearms. Marines use the forearms as a defensive tool to deflect or block attacks. Forearms can also be used as striking weapons to damage or break an opponent's joints and limbs. Marines sustain less self-injury when strikes are conducted with the forearms than when strikes are conducted with fists and fingers.

Elbows. Marines use the elbows as striking weapons. Because of the short distance needed to generate power, elbows are excellent weapons for striking during the close range of close combat.

Legs

The legs are more powerful than any other weapon of the body, and they are less prone to injury when striking. The feet are protected by boots and are the preferred choice for striking.

Feet. Marines use the balls of the feet, the insteps, and the toes to kick an opponent. Marines use the cutting edge of the heels and the heels to stomp on an opponent. Marines must be wearing boots when striking with the toes.

Knees. Like elbows, knees are excellent weapons in the close range of close combat. Knee strikes are most effective while fighting close to an opponent where kicks arc impractical. The opponent's groin area is an ideal target for the knee strike if he is standing upright. Knee strikes can deliver a devastating secondary attack to an opponent's face following an initial attack that caused him to bend at the waist.

3. Target Areas of the Body

During close combat, Marines strive to attack the accessible target areas of an opponent's body. The readily accessible areas will vary with each situation and throughout the engagement. The target areas are divided into five major groups: head, neck, torso, groin, and extremities. The figure below illustrates target areas of the body.

Head

The vulnerable regions of the head are the eyes, temple, nose, ears, and jaw. Massive damage to the head kills an opponent.

Eyes. The eyes are excellent targets because they are soft tissue areas that arc not protected by bone or muscle. Attacks to this area may cause the opponent to protect the area with his hands, allowing Marines to execute a secondary attack to other

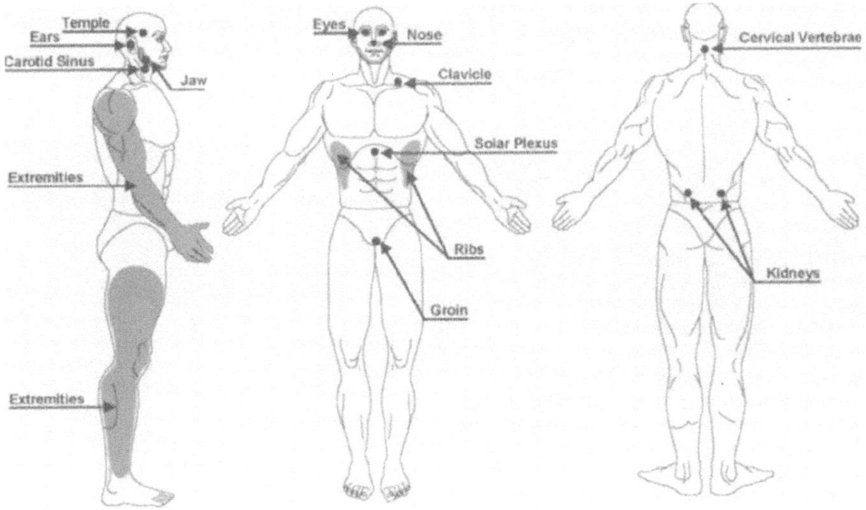

target areas while the opponent uses his hands to protect his eyes.

Temple. The temple is one of the most fragile areas of the head. Powerful strikes to the opponent's temple cause permanent damage and death.

Nose. The nose is very sensitive and easily broken. An attack to this area causes involuntary watering and closing of the opponent's eyes, rendering him vulnerable to secondary attacks. However, through training, individuals can condition themselves to withstand attacks to the nose. Therefore, any attack to the nose must be powerfully delivered.

Ears. Attacks to the ears may cause the eardrum to rupture. But this may not stop or even distract an opponent unless Marines powerfully deliver the strike.

Jaw. The jaw region, when struck forcefully, renders the opponent unconscious. Strikes to the jaw cause painful injuries to the teeth and surrounding tissues (e.g., lips, tongue), but the risk of self-injury is great unless Marines deliver strikes with a hard object such as a helmet, rifle butt, or boot heel.

Neck

The front of the neck, or throat area, is a soft tissue area that is not covered by natural protection. Damage to this region causes the opponent's trachea to swell, closing his airway, which can lead to death.

Carotid Sinus. The carotid sinus is located on both sides of the neck just below the jaw. Strikes to the carotid sinus restrict blood flow to the brain, causing loss of consciousness or death.

Cervical Vertebrae. The cervical vertebrae on the back of the neck, from the base of the skull to the top of the shoulders, contains the spinal cord, which is the nervous system's link to the brain. The weight of the head and the lack of large muscle mass allow damage to the cervical vertebrae and spinal cord. Excessive damage to this area causes pain, paralysis, or death.

Torso

Clavicle. The opponent's clavicle (or collar bone) can be easily fractured, causing immobilization of the arm.

Solar Plexus. Attacks to the opponent's solar plexus or center of the chest can knock the breath out of him and immobilize him.

Ribs. Damage to the opponent's ribs immobilizes him. It may also cause internal trauma.

Kidneys. Powerful attacks to the opponent's kidneys cause immobilization, permanent damage, or death.

Groin

The groin area is another soft tissue area not covered by natural protection. Any damage to this area causes the opponent to involuntarily protect his injured area, usually with his hands or legs. In male opponents, the scrotum is the main target since even a near miss causes severe pain, contraction of the lower abdominal muscles, deterioration of his stance, and possible internal trauma.

Extremities

Rarely will an attack to the opponent's extremities (arms and legs) cause death, but they are important target areas in close combat. Damage to an opponent's joints causes immobilization.

4. Pressure Points of the Body

There are nerves in the human body that, when pressure is applied or when they are struck, allow Marines to control a subject through pain compliance. Marines use pressure points to control an opponent when deadly force is not authorized. They also use pressure points to soften or distract an opponent so a lethal or nonlethal technique can be employed. The figure on page 1-5 illustrates the body's pressure points. Marines execute attacks to pressure points by—

: Rapidly kicking or striking pressure points.
: Slowly applying steady pressure to pressure points.

Infraorbital Nerve

The infraorbital nerve is just below the nose. Marines apply pressure to this nerve with an index finger to control the opponent.

Mastoid Process

The mastoid process is behind the base of the ear and beneath the edge of the jaw. Marines apply inward and upward pressure to this pressure point with the fingers to distract and control the opponent.

Jugular Notch

The jugular notch is at the base of the neck in the notch formed at the center of the clavicle. Marines apply pressure in a quick, stabbing motion with the index finger. Strikes to the jugular notch cause serious damage.

Brachial Plexus (Tie In)

The brachial plexus (tie in) is on the front of the shoulder at the joint. Strikes and pressure applied with the hand are effective on this nerve.

Radial Nerves

Radial nerves are on the inside of the forearms along the radius bones. Strikes and pressure applied with the hand to the radial nerve serve as a softening technique.

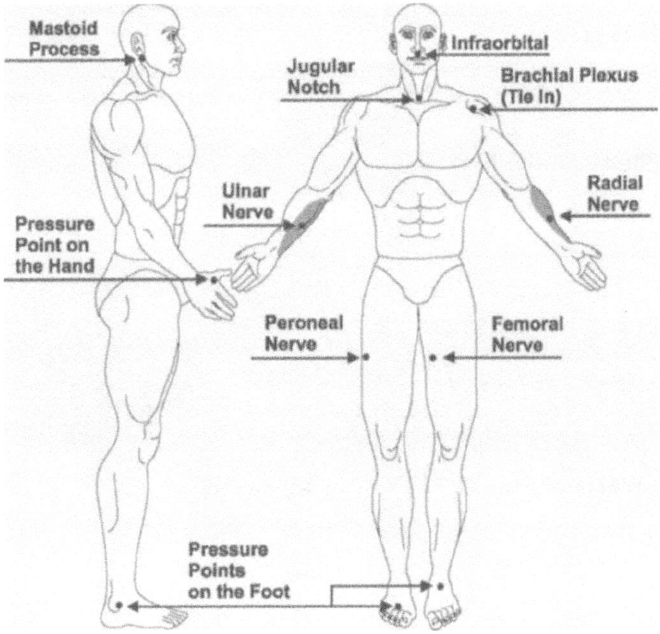

Ulnar Nerve

Ulnar nerves are on the outside of the forearms along the ulnar bones. Strikes and pressure applied with the hand to the ulnar nerve serve as a softening technique.

Pressure Point on the Hand

The hands contain a pressure point on the webbing between the thumbs and index fingers where the two bones of the fingers meet.

To force an opponent to soften or release his grip, Marines apply pressure with their index fingers to this pressure point or strike this pressure point with their fists.

Femoral Nerves

Femoral nerves are on the inside of the thighs along the femur bones. Strikes to the femoral nerve serve as a softening technique.

Peroneal Nerves

Peroneal nerves are on the outside of the thighs along the femur bones. Strikes to the peroneal nerve serve as a softening technique.

Pressure Points on the Feet

There are pressure points on the feet that, when pressure is applied or when they are struck, serve to soften or distract the opponent. Marines apply pressure with the toe, edge, or heel of their boots to the following points:

: The notch below the ball of the ankle.
: The top center of the foot, above the toes.
: The top of the foot where the leg and foot meet.

5. Basic Warrior Stance

Achieving a solid stance is critical to stability and movement throughout any close combat confrontation. The basic warrior stance provides the foundation for stability and movement that is needed to execute close combat techniques. To execute the basic warrior stance, Marines put their feet apart, hands up, elbows in, and chin down.

Feet Apart

Place feet shoulder-width apart.

Keep the head forward and eyes on the opponent, take a half step forward with the left foot, and pivot on the heels so the hips and shoulders are at approximately a 45-degree angle to the right.

Distribute body weight evenly on both legs. Bend the knees slightly.

Hands Up

Curl the fingers naturally into the palm of the hand. Position the thumb across the index and middle fingers. Do not clench the fists. Clenching the fists increases muscular tension in the forearms and decreases speed and reaction time.

Bring the hands up to the face at chin level, with the palms facing each other. Hold the fists up high enough to protect the head, but not so high that they block the field of vision. Ensure continuous eye contact with the opponent.

Elbows In

Tuck the elbows in close to protect the body.

Chin Down

Tuck the chin down to take advantage of the natural protection provided by the shoulders.

6. Angles of Approach and Movement

Marines use movement to control a confrontation and to retain a tactical advantage. Movement increases power and maximizes momentum. By moving around the opponent, Marines gain access to different target areas of the opponent's body. Movement allows Marines to use different weapons of the body and different close combat techniques to attack specific target areas.

Angles of Approach

Marines move anywhere within a 360-degree circle around the opponent to gain a tactical advantage. This circle provides access to different target areas of the opponent's body.

When facing an opponent, Marines move in a 45-degree angle to either side of the opponent. Moving at a 45-degree angle avoids an opponent's strike and puts Marines in the best position to attack the opponent. Marines should avoid being directly in front of an opponent during a confrontation. If a Marine is directly in front of an opponent, the opponent can rely on his forward momentum and linear power to seize the tactical advantage.

Movement

Marines must know how to move in all directions while maintaining the basic warrior stance. During any movement, the legs or feet should not be crossed. Once a movement is completed, the basic warrior stance should be resumed. Maintaining the basic warrior stance protects Marines and puts them in the proper position to launch an attack against an opponent.

Note: Before body movement begins. Marines turn their heads quickly to the new direction. The faster the head turns, the faster the body moves, and the quicker Marines attain visual contact with the opponent.

Forward to the Left. To move forward to the left from the basic warrior stance, Marines-

- Move the left foot forward at a 45-degree angle from the body (approximately 12 to 15 inches), keeping the toe pointed toward the opponent.
- Bring the right foot behind the left foot as soon as the left foot is in place. This returns Marines to the basic warrior stance.

Forward to the Right. To move forward to the right from the basic warrior stance, Marines-

- Move the right foot forward at a 45-degree angle from the body (approximately 12 to 15 inches).
- Bring the left foot, toe pointing toward the opponent, in front of the right foot as soon as the right foot is in place. This returns Marines to the basic warrior stance.

Backward to the Left. To move backward to the left from the basic warrior stance, Marines execute the forward movement in reverse. Marines-

- Move the left foot backward at a 45-degree angle from the body (approximately 12 to 15 inches), keeping the toe pointed toward the opponent.
- Bring the right foot behind the left foot as soon as the left foot is in place. This returns Marines to the basic warrior stance.

Backward to the Right. To move backward to the right from the basic warrior stance, Marines execute the forward movement in reverse. Marines-

- Move the right foot backward at a 45-degree angle from the body (approximately 12 to 15 inches).
- Bring the left foot, toe pointing toward the opponent, in front of the right foot as soon as the right foot is in place. This returns Marines to the basic warrior stance.

7. Balance and Off-Balancing

Balance

In any close combat situation, Marines must strive to maintain balance. The last place to be in a close combat situation is on the ground. Marines must maintain a strong base and a low center of balance, their feet must be a shoulder-width apart, and they must stay on their toes to enable quick movement.

Off-balancing

Marines use off-balancing techniques to control an opponent. These techniques are used to throw an opponent to the ground while Marines remain standing, or they are used to put Marines in a position for an offensive attack.

Off-balancing techniques use the opponent's momentum to move or throw him. For example, if the opponent is charging a Marine, the Marine pulls the opponent to drive him to the ground. Likewise, if the opponent is pulling a Marine, the Marine pushes the opponent to drive him to the ground.

Off-balancing techniques also rely on the power generated by the opponent. For example, during combat a Marine may be tired or outnumbered. Depending on the generated energy and momentum of the opponent, the Marine employs off-balancing techniques with very little effort and still provides effective results.

Because off-balancing techniques rely on the momentum and power generated by the opponent, these techniques are particularly effective for Marines who may be outsized by their opponent or lack their opponent's strength.

Angles of Off-balancing

There are eight angles or directions in which an opponent can be off-balanced: forward, rear, right, left, forward right, forward left, rear right, and rear left.

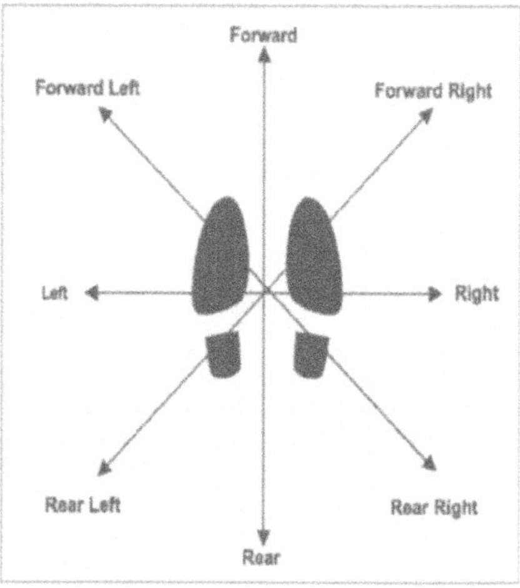

*Note: The angles correspond to the Marine's perspective, not the opponent's. Forward, rear, right, and left are straight angles. Forward right, forward left, rear right, and rear left are considered quadrants that are at a **45**-degree angle in either direction to the front or rear.*

Off-balancing Techniques

Marines off-balance an opponent by pushing, pulling, or bumping the opponent with their hands, arms, or bodies.

To pull, Marines grab an opponent with their hands and drive him forcefully to one of the rear quadrants or to the right or left.

To push, Marines grab the opponent with their hands and drive him forcefully into one of the front quadrants or to the right or left.

Marines execute bumping in the same manner as pushing, but use their shoulders, hips, and legs instead of their hands to off-balance the opponent.

8. Falls

Marines may lose their balance or be thrown to the ground during encounters with an opponent. Marines use falling techniques to absorb the impact of a fall and to quickly return to their feet following an opponent's attack.

Whether falling or being thrown by an opponent, Marines strive to reduce the force of the impact, to prevent serious personal injury, and to increase the chances of survival. Falling techniques use the body's large muscles (back, thighs, buttocks) to protect vital organs and bones from injury and immobilization.

Front Fall

Marines execute a front fall to break a fall on the front. To execute the front fall, Marines-

Bend the elbows and place the palms facing out in a position to spread and absorb the impact of the fall.

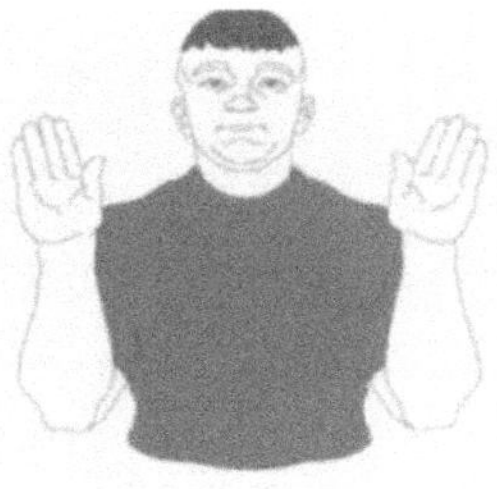

Fall forward, breaking the fall with the forearms and palms. The forearms and hands, down to the fingertips, should strike the ground simultaneous-iy-

Offer resistance with the forearms and hands to keep the head raised off the ground.

Side Fall

Marines execute a side fall to break a fall on the side. To execute the side fall, Marines-

Bring the right arm across the body so the hand is next to the left shoulder with the palm facing inboard.

Fall to the side, breaking the fall with the right arm by slapping the ground and making contact from the shoulder or forearm down to the hand. At the same time, tuck the chin and keep the head raised off the ground. The chin should be tucked to the chest at all times to prevent whiplash.

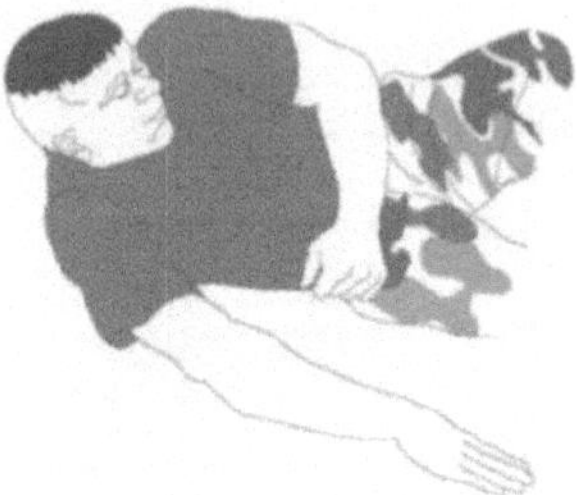

Stretch the right leg out to make contact with the ground and to distribute and absorb the impact. Bend the left leg, allowing the foot to make contact with the ground.

Back Fall

Marines execute a back fall to break the fall when being thrown or falling backward. To execute the backfall, Marines-

Cross the hands in front of the chest and tuck the chin.

Fall backward and slap the ground with the forearms and hands to absorb the impact of the fall and keep the head off the ground.

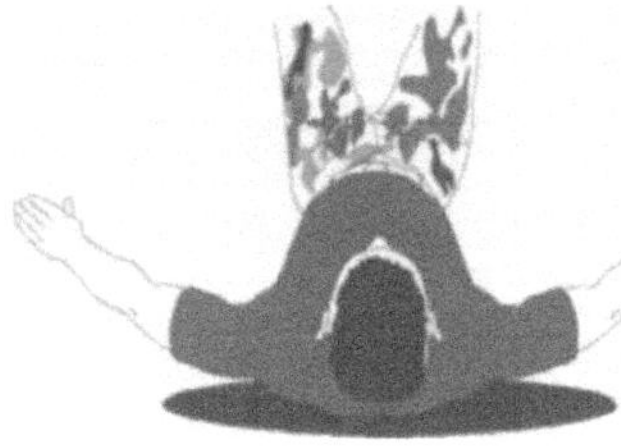

Forward Shoulder Roll

Marines use the forward shoulder roll to break a fall from an opponent's attack and to use the momentum of the fall to get back on their feet quickly. Ideally, Marines execute the forward shoulder roll to a standing position so they can continue fighting. To execute the forward shoulder roll, Marines-

Contact the ground with the back of the right forearm and upper arm. Tuck the chin into the chest.

Roll onto the right shoulder, rolling diagonally across the back to land on the left hip.

Slap the ground with the left arm, absorbing the impact from the shoulder to the hand, palm down.

Keep the left leg straight to absorb as much of the impact as possible. The right leg is bent and the foot hits flat on the ground.

Bend the left leg upon impact to push off with the left knee and leg to a squatting and then a standing position. Forward momentum should carry the Marine to a standing position.

CHAPTER 2

LETHAL AND NONLETHAL WEAPONS TECHNIQUES

This chapter describes all techniques for a right-handed person. However, all techniques can be executed from either side.

In drawings, the Marine is depicted in woodland camouflage utilities; the opponent is depicted without camouflage. In photographs, the Marine is depicted in woodland camouflage utilities; the opponent is depicted in desert camouflage utilities.

1. Bayonet Techniques

WARNING

During training, Marines must have bayonets sheathed. Marines use bayonet dummies to practice bayonet techniques. When practicing offensive and defensive bayonet techniques student-on-student, Marines use pugil sticks.

All Marines armed with a rifle carry a bayonet. The bayonet is an effective weapon if Marines are properly trained in offensive and defensive bayonet techniques. An offensive attack, such as a thrust, is a devastating attack that can quickly end a fight. Defensive techniques, such as the block and parry, can deter the opponent's attack and allow Marines to regain the initiative. Through proper training, Marines develop the courage and confidence required to effectively use a bayonet to protect themselves and destroy the enemy. In situations where friendly and enemy troops are closely mingled and rifle fire and grenades are impractical, the bayonet becomes the weapon of choice.

Holding the Rifle

To execute bayonet techniques, Marines hold the rifle in a modified basic warrior stance. All movement begins and ends with the basic warrior stance. To hold the rifle, Marines—

Use an overhanded grasp to grab the small of the rifle's stock. Use an underhanded grasp to grab the handguards of the rifle.

Lock the buttstock of the rifle against the hip with the right forearm.

Orient the blade end of the rifle toward the opponent.

Offensive Bayonet Techniques

Straight Thrust. Marines use the straight thrust to disable or kill an opponent. It is the most deadly offensive technique because it causes the most trauma to an opponent. Target areas are the opponent' s throat, groin, or face. The opponent's chest and stomach are also excellent target areas if not protected by body armor or combat equipment. To execute the straight thrust, Marines—

Lift the left leg and lunge forward off the ball of the right foot while thrusting the blade end of the weapon forward, directly toward the opponent.

Retract the weapon and return to the basic warrior stance.

Slash. Marines use the slash to kill an opponent or to create an opening in his defense. The target area is the opponent's neck. To execute the slash, Marines—

Extend the left hand back toward the left shoulder.

Thrust the left hand forward and swing it to the right, bring the right hand back toward the hip, and turn the cutting edge of the blade toward the opponent's neck. The movement is a slashing motion so the blade cuts across the opponent's neck.

Horizontal Buttstroke. Marines use the horizontal buttstroke to weaken an opponent's defenses, to cause serious injury, or to set him up for a killing blow. Target areas are the opponent's head, neck, and legs. To execute the horizontal buttstroke, Marines—

Step forward with the right foot and drive the right hand forward. Rotate the hips and shoulders into the strike. Move the left hand back toward the left shoulder.

Strike the opponent with the butt of the weapon.

Vertical Buttstroke. Marines use the vertical buttstroke to weaken an opponent's defenses, to cause serious injury, or to set him up for a killing blow. Target areas are the opponent's groin and face. To execute the vertical buttstroke, Marines—

Step forward with the right foot and drive the right hand straight up.

Pull the left hand back over the left shoulder.

Strike the opponent with the butt of the weapon.

Smash. Marines use the smash as a follow-up technique to the horizontal or vertical buttstroke, particularly if they missed the target. The target area is the opponent's head. To execute the smash following a buttstroke, Marines-

Step forward with the right foot and place the blade end of the weapon over the left shoulder and elevate the right elbow above the shoulders.

Drive the arms straight forward, striking the opponent with the butt of the weapon.

Defensive Bayonet Techniques

Parry. Marines use a parry as a defensive technique to redirect or deflect an attack. A parry is a slight redirection of a linear attack by an opponent; e.g., a straight thrust or a smash. To execute the parry, Marines-

Use the bayonet end of the rifle to redirect the barrel or bayonet of the opponent's weapon.

Lock the weapon against the hip with the right forearm.

Rotate to the right or left, moving the bayonet end of the rifle to parry the opponent's attack. Rotation is generated from the hips.

Redirect or guide the opponent's weapon away from the body by exerting pressure against the opponent' s weapon.

High Block. Marines execute a high block against a vertical attack coming from high to low. To execute the high block, Marines—

Thrust the arms up forcefully at approximately a 45-degree angle from the body. The weapon should be over the top of the head and parallel to the ground. The elbows are bent, but there is enough muscular tension in the arms to absorb the impact and deter the attack.

Low Block. Marines execute the low block

against a vertical attack coming from low to high. To execute the low block, Marines—

Thrust the arms down forcefully at approximately a 45-degree angle from the body. The weapon

should be at or below the waist and parallel to the ground. The elbows are bent, but there is enough muscular tension in the arms to absorb the impact and deter the attack.

Left and Right Block. Marines execute a left or right block against a horizontal buttstroke or a slash. To execute the left or right block, Marines-

Thrust the arms forcefully to the right or left, holding the rifle vertically in the direction of the attack. The elbows are bent, but there is enough muscular tension in the arms to absorb the impact and deter the attack.

Counter Action Following the Block. After deflecting an opponent's attack with a block, Marines counter with a slash or a horizontal buttstroke to regain the initiative. However, the objective in any bayonet fight is to thrust forward with the blade end of the weapon to immediately end the fight.

Group Strategy

On occasion, Marines may engage an opponent as a member of a group or numerous opponents by one's self or as a member of a group. By combining bayonet fighting movements and simple strategies, Marines can effectively overcome their opponent or opponents.

Offensive Strategy: Two Against One. If two bayonet fighters engage one opponent, the fighters advance together.

Fighter 1 engages the opponent while fighter 2 swiftly and aggressively attacks the opponent's exposed flank and destroys the opponent.

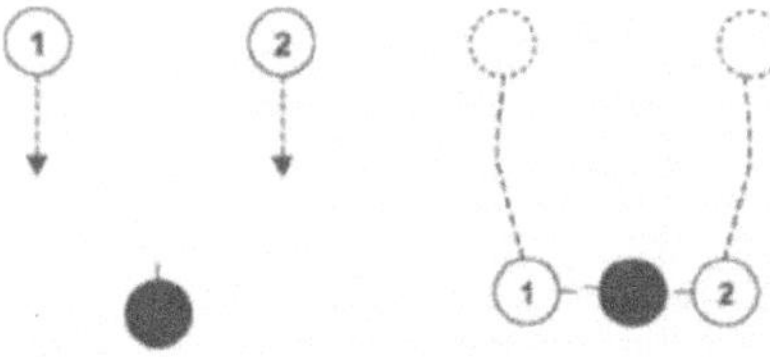

Offensive Strategy: Three Against Two. If three bayonet fighters engage two opponents, the fighters advance together keeping their opponents to the inside.

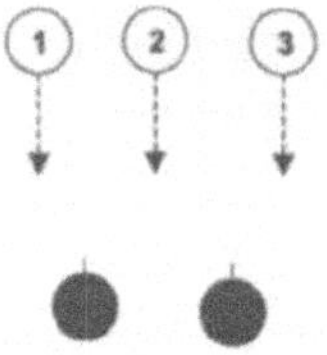

Fighters 1 and 3 engage opponents. Fighter 2 attacks the opponent's exposed flank engaged by fighter 1 and destroys the opponent.

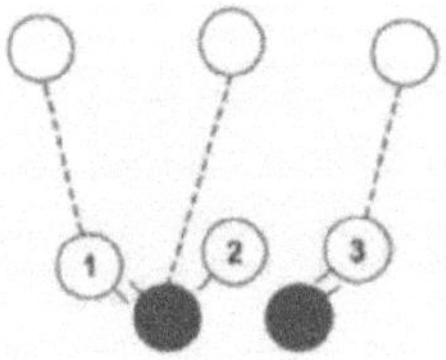

Fighters 1 and 2 turn and attack the exposed flank of the opponent engaged by fighter 3 and destroy the opponent.

Defensive Strategy: One Against Two. If a fighter is attacked by two opponents, the fighter immediately positions himself at the flank of the nearest opponent and keeps that opponent between himself and the other opponent.

Using the first opponent's body as a shield against the second opponent, the fighter destroys the first opponent quickly before the second opponent moves to assist.

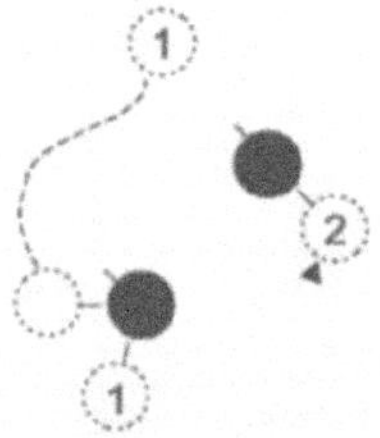

Then, the fighter engages and destroys the second opponent.

Defensive Strategy: Two Against Three. If two fighters are attacked by three opponents, the fighters immediately move to the opponent's flanks.

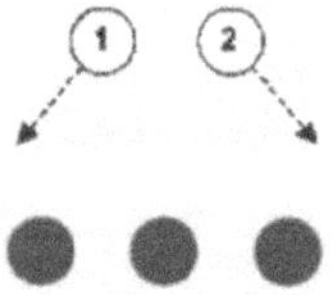

Fighters 1 and 2 quickly attack and destroy their opponents before the third opponent closes in.

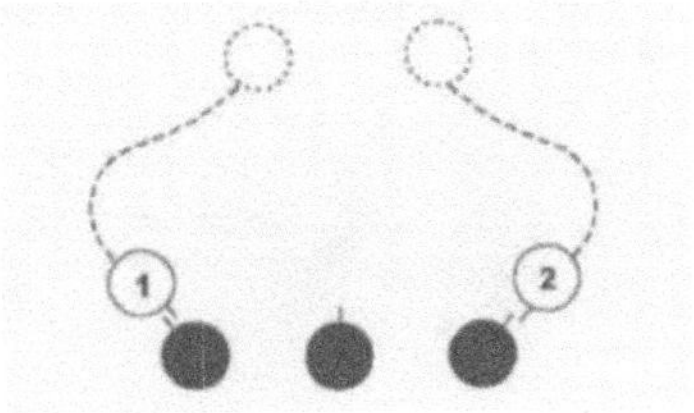

Fighter 1 engages the third opponent while fighter 2 attacks the opponent's exposed flank and destroys the opponent.

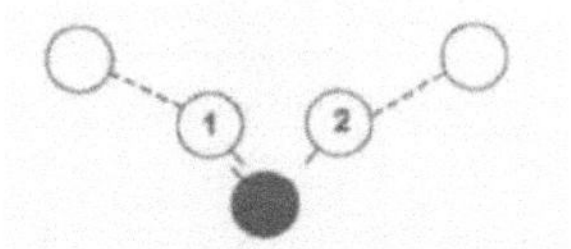

2. Nonlethal Rifle and Shotgun Retention Techniques

Most Marines are armed with the M16A2 service rifle. Marines are taught to keep their weapon with them at all times. Marines must be constantly alert to their surroundings and the people moving in and around their environment. Marines may be confronted by an individual who tries to take theii weapons. If this happens, Marines should not struggle with the individual. To retain positive control of their weapons, Marines must understand and apply weapons retention techniques, otherwise known as armed manipulation. The fol lowing techniques can be used with either the rifle or the shotgun.

Blocking Technique

To execute a blocking technique, Marines-

Stand in a defensive posture.

Use the weapon to block the opponent by thrusting it out firmly, with the elbows bent. Do not try to hit the opponent with the rifle. The rifle is used as a barrier.

Technique if Opponent Grabs Weapon Underhanded

Usually an opponent will try to grab the weapon or block it as an instinctive action. If the opponent uses an underhand grab to seize the handguards of the weapon, Marines—

Trap the opponent's closest finger(s) above the knuckle with the thumb so he cannot release his grip-

Apply bone pressure on the opponent's finger to initiate pain compliance.

Rotate the barrel of the weapon up or down quickly while maintaining pressure on the opponent's hand. At the same time, quickly pivot to off-balance the opponent.

Technique if Opponent Grabs Weapon Overhanded

If the opponent uses an overhand grab to seize the weapon overhanded, Marines—

Trap the opponent's finger to hold his hand in place.

Rotate the barrel to place it across the opponent's forearm and apply downward pressure. This action is similar to an armbar.

Technique if the Opponent Grabs the Muzzle

If the opponent grabs the muzzle of the weapon, Marines—

Rotate the muzzle quickly in a circle motion, Slash downward with the muzzle to release the opponent's grip.

Butt Strikes

Strikes with the butt of the weapon control or ward off an attacker. During any of the retention techniques, Marines use the heel or cutting edge of the weapon to deliver butt strikes to the inside or the outside of the opponent's thigh. The inside butt strike targets the femoral nerve as illustrated in the photo to the left.

The outside butt strike targets the peroneal nerve as illustrated in the photo to the left. Strikes can be made to the outside or inside of the thighs. If a strike to one side of the thigh misses, Marines follow back through with the butt of the weapon on the other side of the thigh.

Off-balancing Techniques

Marines apply off-balancing techniques to throw an opponent to the ground and retain possession of the weapon.

If the opponent grabs the weapon and pushes, Marines should not push on the weapon. They should—

Move with the momentum and movement of the

opponent by pivoting in the direction of the movement by stepping back.

Throw the opponent to the ground with a quick jerking movement by lowering the mu/./le and swinging the butt of the weapon.

If the opponent grabs the weapon and pulls, Marines—

Step on the opponent's foot and push forward to off-balance him and drive him to the ground.

Sweep the opponent's feet out from under him by hooking his leg with the leg and kicking backward.

3. Nonlethal Handgun Retention Techniques

Many Marines are armed with the M9 service pistol. Marines must keep their weapons in their possession at all times. Marines must be constantly alert to their surroundings and the people moving in and around the environment. Marines may be confronted by an individual who tries to take their weapons. To retain positive control of the weapon, Marines must understand and apply handgun retention techniques.

Blocking Technique

Marines perform the following blocking techniques if an opponent attempts to grab their pistol in the holster. Marines—

Place the body between the weapon and the opponent by immediately pivoting so the weapon is away from the opponent.

Step back and away from the opponent while placing the hand on the pistol grip.

Extend the left hand and block, deflect, or strike the opponent's arm.

Armbar Technique

Marines use the armbar technique when an opponent uses his right hand to grab their pistol in the holster. To execute the armbar technique, Marines—

Trap the opponent's right hand by grasping the opponent's wrist or hand with the right hand and applying pressure against the body.

Step back with the right foot and pivot sharply to the right to be next to the opponent. Always pivot in a direction that keeps the weapon away from the opponent.

Straighten the opponent's arm to apply an armbar. The arm should be straight across the torso.

Continue pivoting to the right while pulling back on the opponent's shoulder. This action may break the opponent's arm.

Place the left hand across the opponent's face and apply pressure back and down to take the oppo-

nent to the ground. Pressure applied beneath the nose or on the trachea is equally effective.

Wristlock Technique

Marines use the wristlock technique when an opponent grabs their pistol while it is in the holster with his right hand. To execute the w ristlock technique, Marines—

Grasp the opponent's wrist or hand with the right hand and apply pressure against the body.

Step back and away from the opponent and pivot to the right so the weapon is away from the opponent. Always pivot in a direction that keeps the weapon away from the opponent.

Reach over the opponent's arm with the left hand and grab his hand, applying pressure against his arm with the left forearm. Execute a wristlock.

Incorporate the second hand into the wristlock and, stepping back with the left foot, pivot to the left.

Execute a two-handed wristlock by exerting downward pressure with the thumb and rotating his hand to the left.

Continue to pivot to off-balance the opponent and drive him to the ground.

Note: If the opponent grabs the pistol with his left hand, Marines execute the wristlock with one hand and step in toward the opponent, rather than

away from the opponent.

Softening Techniques

Handgun retention techniques use softening techniques applied to pressure points. Bone pressure

and strikes with the hands (i.e., hammer fist), knees, and feet are also effective softening techniques.

Pressure Points. Marines use pressure point techniques to get the opponent to loosen his grip. Marines use their finger tips to apply pressure to the webbing between the index finger and thumb, the jugular notch, and the brachial plexus tie in. The following figure illustrates pressure applied to the brachial plexus tie in.

Bone Pressure. Bone pressure is the application of pressure on a bone against a hard object to initiate pain compliance. To apply bone pressure, Marines use their hand to trap the opponent's hand on the weapon. Marines apply a slow, steady pressure to the opponent's hand and fingers until his grip is softened or he releases his hold.

Strikes. If it is difficult to apply a retention technique, Marines employ strikes or kicks to force the opponent to loosen his grip. Strikes to the

eyes, the arms (radial nerve), or shoulder (brachial plexus tie in) soften the opponent's grip on the weapon.

Kicks and knee strikes to the peroneal nerve, the femoral nerve, or the groin are effective because the opponent is typically unprepared to counter the strike.

Stomping on the top of the opponent's foot may distract him or loosen his grip on the weapon.

4. Firearm Disarmament Techniques

Marines use firearm disarmament techniques during a close-range confrontation if they are un-

armed and the opponent has a firearm (pistol). These techniques are equally effective if Marines are armed but do not have time to withdraw and present the weapon. The goal of firearm disarmament techniques is to gain control of the situation

so Marines gain the tactical advantage. The goal is not necessarily to get control of the opponent's weapon.

Pistol to the Front

This technique is used when Marines are unarmed and the opponent has a pistol pointing at their front (e.g., head, chest). The technique is the same if the opponent sticks the pistol under the Marine's chin. To execute the counter when an opponent is pointing a pistol toward the front of a Marine, Marines—

Place the hands close to the weapon, about chest high, palms out.

Use the left hand to grab the opponent's forearm and push the opponent's hand with the pistol away to clear the body from in front of the weapon. At the same time, rotate the right shoulder back to clear the body from the weapon.

Maintain control of the opponent's arm.

Grasp the weapon with the right hand by placing the thumb underneath the pistol and the fingers over top of the pistol.

Keep the right hand wrapped tightly around the muzzle and quickly rotate the pistol in the opponent's hand so the muzzle is facing the opponent.

Grasp and pull the opponent's wrist or forearm away from the body while rotating the weapon.

Rotate the weapon toward the opponent while pulling it up and back and out of the opponent's grasp.

Pistol to the Rear

This technique is performed when Marines are unarmed and the opponent has a pistol pointing to the back of the Marine's head. To execute the counter when the opponent is pointing a pistol to the rear, Marines—

Place the hands close together about chest high, palms out.

Step back with the left foot, pivoting on the right foot so the side is against the opponent's front. This action clears the body from the weapon's line of fire. Keep the left hand up.

Pivot on the left foot to face the opponent and, at the same time, raise the left elbow and reach over the top of the opponent's arm with the left arm.

Wrap the left arm tightly around the opponent's arm above his elbow to control it.

Push on the opponent's shoulder with the right hand while pulling up with the left arm to achieve an armbar. This action releases the opponent's grip on the weapon. If necessary, the opponent can be taken to the ground with a leg sweep.

Note: To execute this technique, the weapon must be close to or touching the rear of the Marine. If the weapon is too far away from the body, this technique would be difficult to execute or it would be ineffective.

CHAPTER 3

HAND-HELD WEAPONS

This chapter describes all techniques for a right-handed person. However, all techniques can be executed from either side.

In drawings, the Marine is depicted in woodland camouflage utilities; the opponent is depicted without camouflage. In photographs, the Marine is depicted in woodland camouflage utilities; the opponent is depicted in desert camouflage utilities.

Marines must know how to defend against attacks when an opponent is either unarmed or armed with a held-held weapon. This chapter addresses the combative use of knives, specific weapons of opportunity, and sticks. However, virtually anything can be used as a hand-held weapon.

1. Fundamentals of Knife Fighting

Marines must be trained in knife fighting techniques. Marines experienced in offensive knife techniques can cause enough damage and massive trauma to stop an opponent. When engaged against each other, experienced knife fighters employ various maneuvers and techniques that are specific to knife fighting. Seldom, if ever, will Marines engage an opponent in a classical knife fight.

Note: When armed with a rifle, Marines are issued a bayonet. When armed with a pistol, Marine are issued a combat knife.

Angles of Attack

There are six angles from which an attack with a knife can be launched:

- Vertical strike coming straight down on an opponent.
- Forward diagonal strike coming in at a 45-degree angle to the opponent.
- Reverse diagonal strike coming in at a 45-degree angle to the opponent.
- Forward horizontal strike coming in parallel to the ground.
- Reverse horizontal strike coming in parallel to the ground.
- Forward thrust coming in a straight line to the opponent.

Target Areas of the Body

During any confrontation, the parts of the opponent's body that are exposed or readily accessible will vary. The goal in a knife fight is to attack the body's soft, vital target areas that are readily accessible (e.g, the face, the sides and front of the neck, the lower abdomen [or groin]).

Neck. Carotid arteries, located on either side of the neck, are good target areas because they are not covered by body armor or natural protection.

Lower Abdomen (or Groin). The lower abdomen (or groin region) is a good target area because it is not covered by body armor.

Heart. The heart, if not covered by body armor, is an excellent target which, if struck, can prove fatal in a matter of seconds or minutes.

Secondary Targets. There are secondary target areas that will cause substantial bleeding if an artery is severed. These target areas are not immediately fatal, but can become fatal if left unattended. Attacks to—

The legs can cause a great deal of trauma and prove fatal. For example, the femoral artery located on the inside of the thigh is a large artery which, if cut, will cause extensive blood loss.
The brachial artery, located between the biceps and triceps on the inside of the arm, can cause extensive bleeding and damage.
The arm's radial and ulnar nerves can cause extensive bleeding and damage.

Movement

Marines can move anywhere within a 360-degree circle around the opponent. This allows accessibility to different target areas of the opponent's body. Marines should avoid being directly in front of an opponent because the opponent can rely on his forward momentum to seize the tactical advantage. If Marines face an opponent, movement is made in a 45-degree angle to either side of the opponent. This angle avoids an opponent's strike and places Marines in the best position to attack an opponent.

Wearing the Combat Knife

Marines must wear the combat knife where it is easily accessible and where it can best be retained. It is recommended the combat knife be worn on the weak side hip, blade down, sharp edge facing forward. Marines can place it behind the magazine pouch where it is easily accessible to them, but not easily grabbed by an opponent.

Grip

The grip on the knife should be natural. Marines grasp the knife's grip with the fingers wrapped around the grip naturally as it is pulled out of its sheath. This is commonly known as a hammer grip. The blade end of the knife is always facing the opponent.

Stance

Marines use the basic warrior stance as the foundation for knife techniques. The left hand is a vertical shield that protects either the ribs or the head and neck. The right elbow is bent with the blade pointing forward toward the opponent' s head. This position serves as an index point, where all techniques are initiated.

Principles of Knife Fighting

The following are key principles of knife fighting:

Execute movements with the knife blade within a box, shoulder-width across from the neck down to the waistline. The opponent has a greater chance of blocking an attack if the blade is brought in a wide, sweeping movement to the opponent.
Close with the opponent, coming straight to the target.
Move with the knife in straight lines.
Point the knife's blade tip forward and toward the opponent.
Apply full body weight and power in each of the knife techniques. Full body weight should be put into the attack in the direction of the blade's movement (slash or thrust).
Apply constant forward pressure with the body and blade to keep the opponent off-balanced.

2. Knife Fighting Techniques

Slashing Techniques

Marines use slashing techniques to close with an enemy. Slashing techniques distract the opponent or damage the opponent so Marines can close in. Typically, Marines target the opponent's limbs, but any portion of the body that is presented can become a target.

Vertical Slash Technique. The vertical slash follows a vertical line straight down through the target. To execute the vertical slash, Marines—

Thrust the right hand out and bring the weapon straight down on the opponent, continuing to drag the knife down through the opponent's body.

Maintain contact on the opponent's body with the blade of the knife.

Forward Slash Technique. The forward slash follows a straight line in a forehand stroke, across the target areas of either the neck (high diagonal slash) or abdominal region (low horizontal slash). To execute the forward slash, Marines-
Extend the right hand while simultaneously rotating the palm up until the knife blade makes contact with the opponent.

Snap or rotate the wrist through the slashing motion to maximize blade's contact with the opponent.

Drag the knife across the opponent's body, from right to left, in a forehand stroke. The movement ends with the forearm against the body and the knife at the left hip with its blade oriented toward the opponent.

Reverse Slash Technique. The reverse slash is a follow-up technique to a forward attack. It allows Marines both a secondary attack and the ability to resume the basic warrior stance. The reverse slash follows a straight line in a backhand stroke, across the target areas of either the neck (high diagonal slash) or abdominal region (low horizontal slash). To execute the reverse slash, Marines—

Extend the right hand while simultaneously rotating the palm down until the knife blade makes contact with the opponent.

Snap or rotate the wrist through the slashing motion to maximize the blade's contact with the opponent.

Drag the knife across the opponent's body, from left to right, in a backhand stroke. Maintain contact on the opponent's body with the blade of the knife.

Thrusting Techniques

The primary objective of knife fighting is to insert the blade into an opponent to cause extensive damage and trauma. This is done with a thrusting technique. Thrusting techniques are more effective than slashing techniques because of the damage they can inflict. However, Marines use slashing techniques to close with the enemy so that they are closer to the opponent, which allows them to use the thrusting technique.

Vertical Thrust. The thrusting motion follows a vertical line straight up through the target (low into the abdomen region or high into the neck). To execute the vertical thrust, Marines—

Thrust the right hand toward the target, inserting the knife blade straight into the opponent.

Pull the knife out of the opponent.

Forward Thrust. The forward thrust follows a straight line straight into the opponent's neck (high thrust) or abdominal region (low thrust). To execute the forward thrust, Marines—

Thrust the right hand, palm down, toward the target, inserting the knife blade straight into the opponent.

Rotate the palm up once the knife is inserted to twist the blade.

Drop the right elbow and bring the knife to the opposite side of the opponent's body from where it was inserted. At the same time, rotate the hips and shoulders downward to bring body weight to bear on the attack.

Reverse Thrust. The reverse thrust is a follow-up technique to a forward attack. It allows Marines both a secondary attack and the ability to resume the basic warrior stance. The reverse thrust follows a horizontal line straight into the opponent's neck (high thrust) or abdominal region (low thrust). To execute the reverse thrust, Marines-

Bend the right arm, crossing the arm to the left side of the body.

Thrust the right hand, palm up, toward the target, inserting the knife blade straight into the opponent.

Rotate the palm down to twist the blade once the knife is inserted.

Bring the knife to the opposite side of the opponent's body from where it was inserted.

3. Weapons of Opportunity

During an unarmed close combat situation, Marines use their bodies as weapons, but they should be ready and able to use anything around them as a weapon. For example, Marines could throw sand or liquid in an opponent's eyes to temporarily impair his vision or smash the opponent's head with a rock or helmet. Marines must use whatever means are available and do whatever it takes to take control of the situation and to win, or they face the possibility of losing their lives. Some weapons of opportunity are discussed in the following subparagraphs.

Entrenching Tool

An entrenching tool (E-tool) is commonly carried by Marines. It can be an excellent weapon, especially when sharpened. Marines can use the E-tool to block, slash, and thrust at an opponent.

Tent Pole and Pins

Marines can use tent poles and pins to block, strike, or thrust at an opponent.

Web Belt

Marines can stretch a web belt between their hands to block attacks by an opponent.

Battlefield Debris

Marines can use debris on the battlefield (e.g., sticks, glass, a sharp piece of metal) to cut, slash, or stab an opponent. They can also use other types of debris such as shovel or ax handles, boards, metal pipes, or broken rifles to strike an opponent or apply a choke.

Helmet

A helmet can be used to strike an opponent on an unprotected area like the head and face. Grasp the rim of the helmet and thrust the arms forward, striking the opponent with the top of the helmet.

4. Fundamentals of Combative Stick

On the battlefield, Marines must be ready and able to use anything as a weapon. They must learn and be able to use techniques that can be employed with most weapons of opportunity. Among these techniques are combative stick techniques. Combative stick techniques can be used with a stick, a club, a broken rifle, an E-tool, or even a web belt.

Angles of Attack

There are six angles from which an attack with a hand-held weapon can be launched:

- Vertical strike coming straight down on an opponent.
- Forward diagonal strike coming in at a 45-degree angle to the opponent.
- Reverse diagonal strike coming in at a 45-degree angle to the opponent.
- Forward horizontal strike coming in parallel to the ground.
- Reverse horizontal strike coming in parallel to the ground.
- Forward thrust coming in a straight line to the opponent.

Grip

Grasp the stick about 2 inches from its base.

Stance

The basic warrior stance serves as the foundation for combative stick techniques. The left hand becomes a vertical shield that protects the ribs or the head and neck. Depending on how heavy the weapon is, it should be held at a level approximately shoulder height.

Movement

Movement during combative stick techniques is the same as it is for other close combat techniques. Marines can move anywhere within a 360-degree circle around the opponent. This allows accessibility to different target areas of the opponent's body and gains a tactical advantage. Marines should avoid being directly in front of an opponent because the opponent can rely on his forward momentum to seize the tactical advantage. If Marines face an opponent, movement is made in a 45-degree angle to either side of the opponent. This angle avoids an opponent's strike and places Marines in the best position to attack an opponent.

5. Combative Stick Techniques

Strikes

Strikes are intended to inflict as much damage on an opponent as possible. Striking techniques apply to a weapon of opportunity such as a stick, a tent pole, a club, a broken rifle, an E-tool, or a Pipe-

Vertical Strike. To execute the vertical strike, Marines—

Bend the right arm, extending the weapon over the back of the right shoulder.

Rotate the forearm straight down off the elbow to bring the weapon down on the opponent.

Rotate the hips and shoulders forcefully toward the opponent.

Follow through with the strike by allowing the weight of the weapon to go through the target area of the body.

Forward Strike. To execute a forward strike, Marines—

Step forward with the left foot in the direction of the strike.

Bend the right arm with the elbow extending out to the right and the weapon extended over the right shoulder.

Rotate the forearm to the right of the elbow to bring the weapon down onto the opponent. At the same time, forcefully rotate the hips and shoulders toward the opponent.

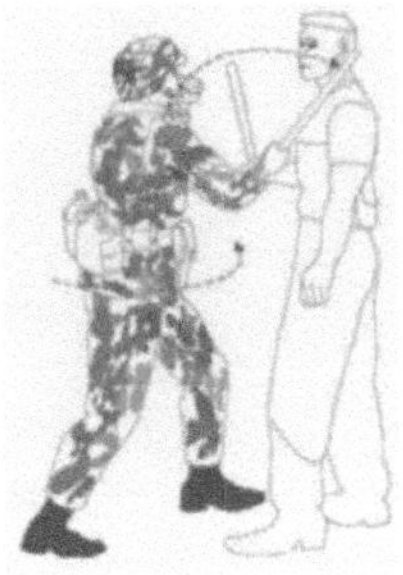

Follow through with the strike by allowing the weight of the weapon to go through the target area of the body.

Reverse Strike. The reverse strike is a follow-up technique to a forward strike. It allows Marines both a secondary attack and the ability to resume the basic warrior stance.To execute a reverse strike, Marines—

Step forward with the right foot in the direction of the strike.

Bend the right arm with the hand near the left shoulder. The weapon is extended over the left shoulder.

Rotate the forearm to the right of the elbow to bring the weapon down onto the opponent. At the same time, forcefully rotate the hips and shoulders toward the opponent.

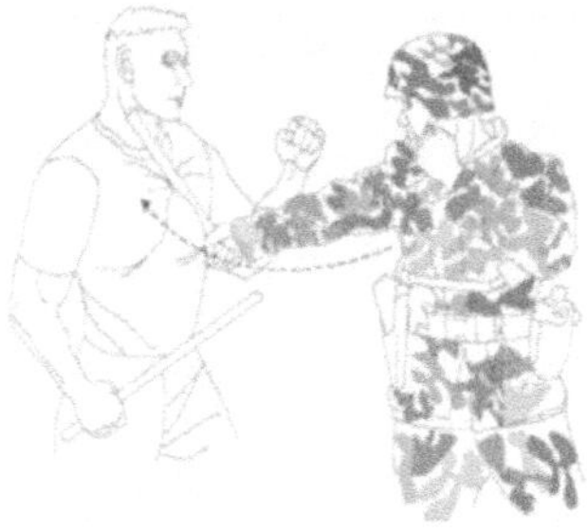

Follow through with the strike by allowing the weight of the weapon to go through the target area of the body.

Forward Thrust. To execute the forward thrust, Marines—

Grasp the stick with the left hand, palm up, in a position where the stick can be controlled with two hands.

Lift the left leg and lunge forward off the ball of the right foot. At the same time, thrust the end of the weapon directly toward the opponent by thrusting both hands forward in a straight line.

6. Blocking Techniques

A block is meant to deter or deflect an attack by an opponent. A block sets up Marines for a follow-on attack against the opponent. Blocks are executed by deflecting, rather than hitting or following through like a strike.

Blocks Against Unarmed Attacks

To block against an unarmed attack, Marines—

Step forward at a 45-degree angle with either the right or left foot. This moves the body out of the line of attack.

Raise the left arm and block the strike with the meaty portion of the forearm.

Employ the stick by one of two techniques:

Strike the opponent with the stick.
Use the stick to block on two points of contact. When the stick is used to block, it serves as an incidental strike. This technique is only used if Marines have closed with the opponent and are inside his strike.

Blocks Against Armed Attacks

Block for a Vertical Strike. To execute the block for a vertical strike, Marines—

Step forward with the left foot at a 45-degree angle to the left. This moves the body out of the line of attack.

Block on two points of contact to disperse the impact of the attack:

Block the opponent's stick by positioning the stick so it is perpendicular to the opponent's stick. If the stick is not perpendicular to the opponent's stick, the stick can slide through and make contact on the Marine.
Block the opponent's wrist or forearm with the back of the left forearm.

Note: Use the stick to block the opponent's arm if closer to the opponent. It is the same movement as blocking with the arm, except the opponent's arm is blocked with both the stick and the arm.

Block for a Forward Strike. To execute the block for a forward strike, Marines-

Step forward with the left foot at a 45-degree angle to the left. This moves the body out of the line of attack and inside the opponent's strike.

Block on two points of contact to disperse the impact of the attack:

- Block the opponent's wrist or forearm with the meaty portion of the back of the left forearm.
- Strike the opponent's attacking biceps with the stick.

Block for a Reverse Strike. To execute the block for a reverse strike, Marines—

Step forward at a 45-degree angle with the right foot to the right. This moves the body out of the line of attack and inside the opponent's strike.

Block on two points of contact to disperse the impact of the attack:

> Block the opponent's stick by positioning the stick so it is perpendicular to the opponent's stick.

Block the opponent's forearm with the meaty portion of the left forearm.

Note: If closer to the opponent, block his triceps with the back of the left forearm and strike his forearm with the stick.

7. Unarmed Against Hand-Held Weapons

If Marines are engaged against an opponent with a knife, a stick, or some other weapon of opportunity, they must establish and maintain an offensive mindset, not a defensive mindset. Their survival depends on it. They cannot afford to think about getting cut or hurt.

Angles of Attack

Before Marines can learn to block or counter an attack with a hand-held weapon (e.g., knife, stick), they must know from what angle the opponent is attacking. There are six angles from which an opponent will typically attack with a hand-held weapon:

- Vertical strike coming in straight down on the Marine.
- Forward diagonal strike coming in at a 45-degree angle to the Marine.
- Reverse diagonal strike coming in at a 45-degree angle to the Marine.
- Forward horizontal strike coming in parallel to the ground.
- Reverse horizontal strike coming in parallel to the ground.
- Forward thrust coming in a straight line to the Marine.

Blocks

If the opponent has a hand-held weapon (e.g., knife or stick), Marines parry the opponent's hand or arm to block the attack.

Basic Block. To execute the basic block technique, Marines-

Step forward at a 45-degree angle to move out of the line of the attack. Always step in the direction of the strike.

Thrust the forearms forward, hands up, against the opponent's attacking arm. Contact is made on the opponent's arm with the backs of the forearms.

Block for a Vertical Strike. To execute the block against a vertical strike, Marines—

Step forward with the left foot at a 45-degree angle to the left to move out of the line of attack.

Thrust the forearms forward, hands up, against the outside of the opponent's attacking arm.

Block for a Forward Strike. To execute the block against a forward diagonal or forward horizontal strike, Marines-

Step forward with the left fool inside the opponent's attacking arm.

Block the attack with both arms bent so the forearms make contact with the opponent's biceps and forearm.

Block for a Reverse Strike. To execute the block against a reverse diagonal or reverse horizontal strike, Marines—

Step forward with the right foot to the outside of the opponent's attacking arm.

Block the attack with both arms bent so the forearms make contact with the opponent's triceps and forearm.

Block for a Forward Thrust. To execute the block against a forward thrust, Marines—

Bend at the waist, move the hips backward, and jump backward with both feet to move away from the attack. This action is known as "hollowing out."

Hollow out and block the attack with the arms bent and hands together on top of the attacking arm.

Overlap the hands slightly so one thumb is on top of the other hand's index finger. The other thumb should be under the other hand's index finger.

8. Counters to Hand-Held Weapon Attacks

Principles of Counters

A counter is used to control the situation to regain the tactical advantage and end the fight. Regardless of the type of weapon or angle of attack, the following principles apply to countering the attack with a hand-held weapon:

Move out of the line of attack. Movement is executed in a 45-degree angle forward to the left or right.

- Block the attack.

Note: The first two actions are taken simultaneously.

- Take control of the weapon by controlling the hand or arm that is holding the weapon. Never attempt to grab the opponent's weapon.
- Execute the appropriate follow-up to end the fight; e.g., strikes, joint manipulations, throws, or takedowns (see chap. 8). Marines should continue to attack the opponent until the fight ends.

Counter Techniques

There are two techniques that can be used to counter any armed attack: forward armbar counter and reverse armbar counter. These techniques can be used to counter a vertical attack, a forward diagonal strike, or a forward horizontal strike. With minor variations, the same techniques are used to counter reverse strikes. A third technique, the bent armbar counter, is used to counter a vertical attack.

Forward Armbar Counter. To execute the forward armbar counter to an attack coming from a forward strike, Marines-

Step forward with the left foot inside the opponent' s attacking arm.

Block the attack with both arms bent so the forearms make contact with the opponent's biceps and forearm.

Slide the left arm over the opponent's forearm and wrap the arm tightly around his arm, trapping the opponent's attacking arm between the biceps and torso.

Place the right hand on the opponent's shoulder or upper arm to further control his arm and to effect an armbar.

Execute an armbar and continue to exert steady pressure against the arm.

Reverse Armbar Counter. To execute the reverse armbar counter to an attack coming from a forward strike, Marines-

Step forward with the left foot inside the opponent' s attacking arm.

Block the attack with both arms bent so the forearms make contact with the opponent's biceps and forearm.

Control the opponent's arm with the left hand and pivot to the right so the back is against the opponent's side. Immediately slide the right arm over the opponent's biceps and wrap the arm tightly around his arm, trapping the opponent's attacking arm between the biceps and torso.

Grasp the opponent's wrist with the left hand and twist his thumb away from the body.

Take the opponent to the ground with an armbar takedown if he does not drop the weapon from the wristlock.

Bent Armbar Counter. This counter is particularly effective against a vertical attack. To execute the bent armbar counter, Marines—

Step forward with the left foot inside the opponent's attacking arm.

Block the attack with both arms bent so the forearms make contact with the opponent's biceps and forearm.

Grasp the opponent's forearm with the left hand. At the same time, slide the right arm underneath the opponent's triceps and grasp the opponent's forearm or wrist with the right hand.

Apply downward pressure with the hands against the opponent's forearm to off-balance the opponent.

CHAPTER 4

STRIKES

This chapter describes all techniques for a right-lmnded person. However, all techniques can be executed from either side.

In drawings, the Marine is depicted in woodland camouflage utilities; the opponent is depicted without camouflage. In photographs, the Marine is depicted in woodland camouflage utilities; the opponent is depicted in desert camouflage utilities.

Strikes are unarmed individual hitting techniques. Strikes use the hands, elbows, knees, feet, and, in some instances, other parts of the body as personal weapons. Marines must know how to execute strikes effectively. They must also know how to counter strikes from an opponent.

1. Principles of Punches

Muscle Relaxation

Muscle relaxation is crucial when executing punches. The natural tendency in a fight is to tense up, which results in rapid fatigue and decreased power generation. Marines who remain relaxed during a close combat situation generate greater speed, which results in greater generation of power. Relaxing the forearms generates speed and improves reaction time. At the point of impact, Marines clench the fist to cause damage to the opponent and avoid injury to the wrist and hand.

Weight Transfer

Weight transfer is necessary to generate power in a punch. Marines accomplish this by-

Rotating their hips and shoulders into the attack.

Moving their body mass straight forward or backward in a straight line.

Dropping their body weight into an opponent. Body mass can be transferred into an attack from high to low or from low to high.

Rapid Retraction

When Marines deliver a punch, rapid retraction of the fist is important. Once the hand has made contact with the target, Marines quickly return to the basic warrior stance. Rapid retraction-

Returns the hand and arm to the protection afforded by the basic warrior stance.

Prevents the opponent from being able to grab the hand or arm.

Permits the hand and arm to be "chambered" or "re-cocked" in preparation for delivering a subsequent punch.

Telegraphing

Telegraphing a strike occurs when body movements inform the opponent of the intention to launch an attack. Staying relaxed helps to reduce telegraphing.

Often, an untrained fighter telegraphs his intention to attack by drawing his hand back in view of his opponent, changing facial expression, tensing neck muscles, or twitching. These movements, however small, immediately indicate an attack is about to be delivered. If the opponent is a trained fighter, he may be able to evade or counter the attack. If the opponent is an untrained fighter, he may still be able to minimize the effect of an attack.

2. Punches

Punches may be thrown during any hand-to-hand confrontation. Most people resort to punching because it is a natural reaction to a threat. The purpose of a punch is to stun the opponent or to set him up for a follow-up finishing technique. However, punches should only be executed to the soft tissue areas of an opponent. A correctly delivered punch maximizes the damage to an opponent while minimizing the risk of injury to Marines.

Basic Fist

Punches are executed using the basic fist. To make the basic fist, the fingers are curled naturally into the palm of the hand and the thumb is placed across the index and middle fingers. Do not clench the fist until movement has begun. This reduces muscular tension in the forearm and increases speed and reaction time. Just before impact, Marines exert muscular tension on the hand and forearm to maximize damage to the opponent and reduce their chances of injury. Contact should be made with the knuckles of the index and middle fingers.

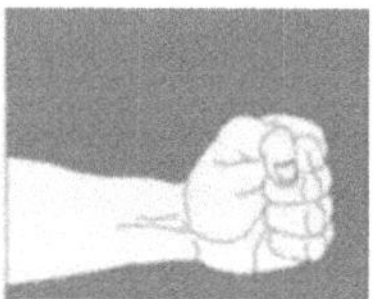

Picture shows finger position only.

When striking with the basic fist, Marines must keep the hand straight, or in line, with the wrist to avoid injury to the wrist.

Lead Hand Punch

The lead hand punch is a snapping straight punch executed by the forward or lead hand. It is a fast punch designed to keep the opponent away and to set up an attack. A lead hand punch conceals movement and allows Marines to get close to the opponent. Lead hand punches should strike soft tissue areas, if possible. To execute the lead hand punch, Marines-

Snap the lead hand out to nearly full extension, while rotating the palm to the ground.

Contact the opponent with the first two knuckles of the fist.

Retract the hand immediately, resuming the basic warrior stance.

Rear Hand Punch

The rear hand punch is a snapping punch executed by the rear (right) hand. It is a power punch designed to inflict maximum damage on the opponent. Its power comes from pushing off the rear leg and rotating the hips and shoulders. To execute the rear hand punch, Marines—

Rotate the hips and shoulders forcefully toward the opponent and thrust the rear hand straight out, rotating the palm down, to nearly full extension.

Shift body weight to the lead foot while pushing off on the ball of the rear foot.

Contact the opponent with the first two knuckles of the fist.

Retract the hand immediately.

Uppercut

The uppercut is a powerful punch originating below the opponent's line of vision. It is executed in an upward motion traveling up the centerline of the opponent's body. It is delivered in close and usually follows a preparatory strike that leaves the target area unprotected. When delivered to the chin or jaw, the uppercut can render an opponent unconscious, cause extensive damage to the neck, or sever the tongue. To execute the uppercut, Marines—

Bend the arms, rotating the palm inboard. The distance the arms bend depends on how close the opponent is.

Rotate the hips and shoulders forcefully toward the opponent, thrusting the fist straight up toward the opponent's chin or jaw.

Contact the opponent with the first two knuckles of the fist.

Retract the hand immediately.

Hook

The hook is a powerful punch that is executed close in and is usually preceded by a preparatory strike. To execute the hook, Marines—

Thrust the right arm in a hooking motion toward the opponent, keeping the elbow bent while forcefully rotating the right shoulder and hip toward the opponent.

Contact the opponent with the first two knuckles of the fist. Continue rotating the shoulder and hip, following through with the fist to the target.

Retract the hand immediately.

3. Strikes with the Upper Body

Strikes stun the opponent or set him up for a follow-up finishing technique. The hands, forearms, and elbows are individual weapons of the arms that can be used to execute strikes including the hammer fist, chin jab, knife hand, eye gouge, and elbow strikes. These strikes provide a variety of techniques that can be used in any type of close combat situation.

Principles of Execution

There are several principles of execution that ensure a strike's effectiveness.

Generating Power. Weight transfer is necessary to generate maximum power in a punch. Marines accomplish this by-

- Rotating the hips and shoulders into the attack.
- Moving the body mass straight forward or backward in a straight line.
- Dropping body weight into an opponent. Body mass can be transferred into an attack from high to low or from low to high.

Muscular Tension. The arms are relaxed until the moment of impact. At the point of impact, Marines apply muscular tension in the hand and forearm to maximize damage to the opponent and to avoid injury to the hand. The arms are relaxed until the moment of impact.

Hit and Stick and Follow-Through. A strike should be delivered so that the weapon (e.g., hand, elbow) hits and remains on the impact site (target) and follows through the target. This technique inflicts maximum damage on the opponent.

Strikes with the arms are executed with "heavy hands"; i.e., the strike is executed by driving through with the strike to allow the weight of the hand to go through the target area of the body. Contact with the opponent should be made with the arm slightly bent, and the arm extends as it moves through the target. This technique allows Marines to deliver strikes effectively without executing full force.

Movement. Movement puts Marines in the proper position for launching an attack against an opponent as well as providing protection.

Movement is initiated from the basic warrior stance and ends by resuming the basic warrior stance. Strikes can be performed with either the left or right arm depending upon—

The angle of attack.
The position of the opponent.
The opponent's available, vulnerable target areas.

Target Areas of the Body

For each strike, there are target areas of the body that, when struck, maximize damage to an opponent. Strikes use gross motor skills as opposed to fine motor skills. The target areas of the body are just that—areas. Pinpoint accuracy on a specific nerve is not needed for the strike to be effective.

Hammer Fist

Striking with the hammer fist concentrates power in a small part of the hand, which, when transferred to the target, can have a devastating effect. The striking surface of the hammer fist is the meaty portion of the hand below the little finger. To execute the hammer fist strike, Marines-

Make a fist and bend the arm at approximately a 45- to 90-degree angle. At the same time, rotate the right hip and right shoulder backward.

Thrust the fist forward onto the opponent while rotating the right hip and shoulder forward.

Rotate the wrist so the hammer fist makes contact on the opponent.

Chin Jab

The chin jab can immediately render an opponent unconscious and cause extensive damage to the neck and spine. The striking surface is the heel of the palm of the hand. To execute the chin jab, Marines—

Bend the right wrist back at a 90-degree angle with the palm facing the opponent and the fingers pointing up.

Keep the right arm bent and close to the body. Extend the hand into a concave position with the fingers slightly spread apart.

Step forward with the left foot toward the opponent, keeping the feet approximately shoulder-width apart and the knees bent. This is done to close with the opponent.

Keep the right arm bent and close to the side. Thrust the palm of the hand directly up under the opponent's chin. At the same time, rotate the right hip forward to drive the body weight into the attack to increase the power of the strike. The attack should travel up the centerline of the opponent's chest to his chin.

Knife Hand

The knife hand is one of the most versatile and devastating strikes. The striking surface is the cutting edge of the hand, which is the meaty portion of the hand below the little finger extending to the top of the wrist. The striking surface is narrow, allowing strikes on the neck between the opponent's body armor and helmet. The knife hand strike is executed from one of three angles: outside, inside, and vertical.

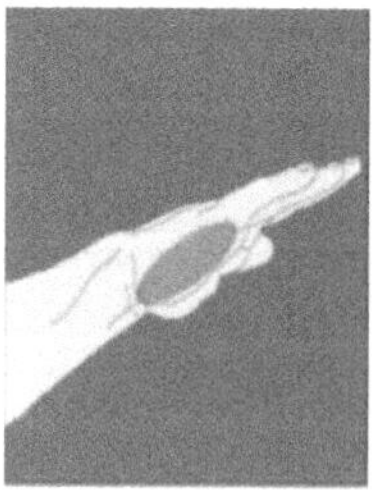

Outside Knife Hand. To execute the outside knife hand strike, Marines—

Execute a knife hand by extending and joining the fingers of the right hand and placing the thumb next to the forefinger (like saluting).

Retract the right hand. At the same time, rotate the right hip and right shoulder backward.

Thrust the knife hand forward (horizontally) onto the opponent while rotating the right hip and shoulder forward.

Inside Knife Hand. To execute the inside knife hand strike, Marines—

Execute a knife hand.

Bring the right hand over the left shoulder. At the same time, rotate the right shoulder forward and the left hip forward.

Thrust the knife hand forward (horizontally) onto the opponent while rotating the right hip and shoulder forward and the left shoulder backward.

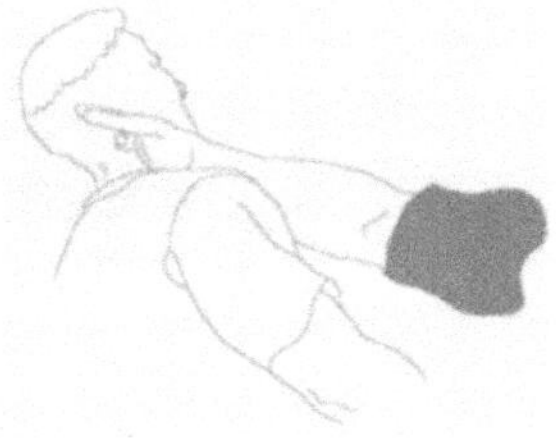

Vertical Knife Hand. When thrown vertically, the knife hand strike comes straight down in a straight line.

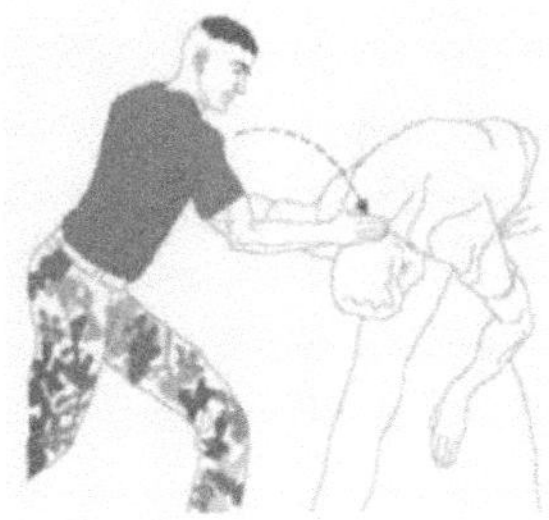

Eye Gouge

The eye gouge is used to attack an opponent's eyes, blinding him so follow-up strikes can be executed. The striking surface is the tips of the fingers or thumb. To execute the eye gouge, Marines—

Extend the right hand with the fingers slightly spread apart to allow entry into the eye sockets.

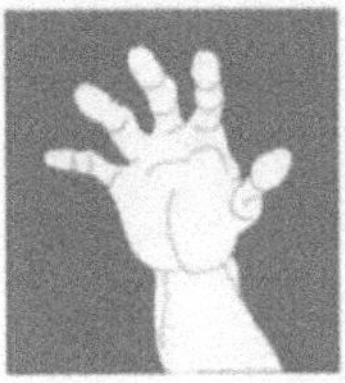

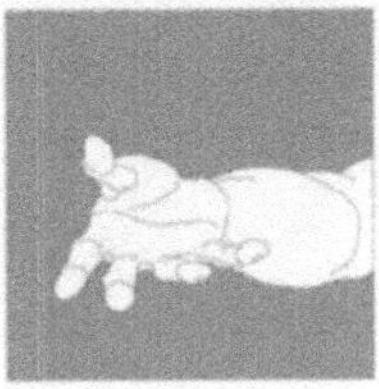

Place the palm of the hand either toward the ground or toward the sky and thrust the right hand forward into the opponent's eyes.

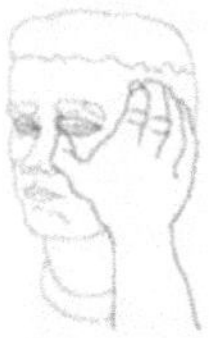

Thrust the hand forward at the opponent's nose level so the fingers or thumb slide naturally into the grooves of the opponent's eye sockets.

Elbow Strikes

The elbow is a powerful weapon that can be used in several different ways to attack virtually any part of an opponent's body. Elbow strikes can be performed either vertically (upward or downward) or horizontally (forward or reverse). The striking surface is 2 inches above or below the point of the elbow, depending upon the angle of attack, the opponent's attack angle, and the position of the opponent.

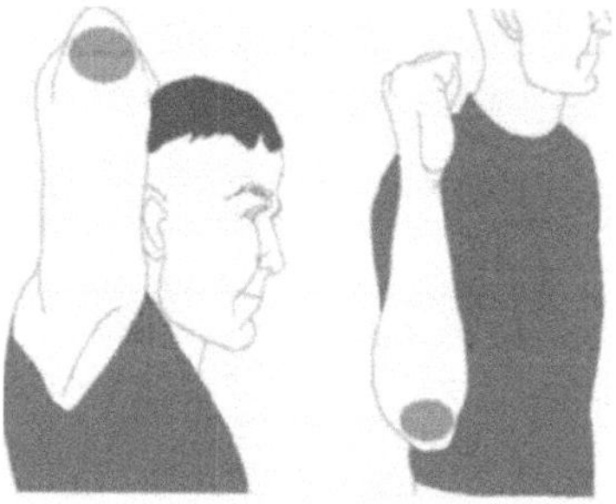

Above Elbow Below Elbow

Vertical Elbow Strike (Up). To execute an upward vertical elbow strike, Marines-

Bend the right elbow, keeping the fist close to the body. The fist is at shoulder level and the elbow is next to the torso.

Thrust the elbow vertically upward toward the opponent while rotating the right shoulder and hip forward to generate additional power.

Contact the opponent with the right forearm 2 inches above the point of the elbow.

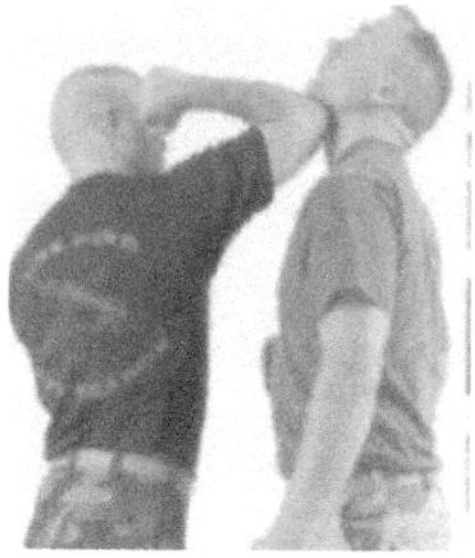

Vertical Elbow Strike (Down). To execute a downward vertical elbow strike, Marines—

Bend the right elbow, keeping the fist close to the body. The fist is on the shoulder and the elbow is raised well above the shoulder.

Thrust the elbow vertically downward toward the opponent while dropping body weight into the attack to generate additional power.

Contact the opponent with the right triceps 2 inches above the point of the elbow.

Horizontal Elbow Strike (Forward). To execute the forward horizontal elbow strike, Marines-

Tuck the right fist near the chest with the palm heel facing the ground.

Thrust the right elbow horizontally forward toward the opponent. The forearm is parallel to the ground.

Rotate the right shoulder and hip forward to generate additional power.

Contact the opponent with the right forearm 2 inches below the point of the elbow.

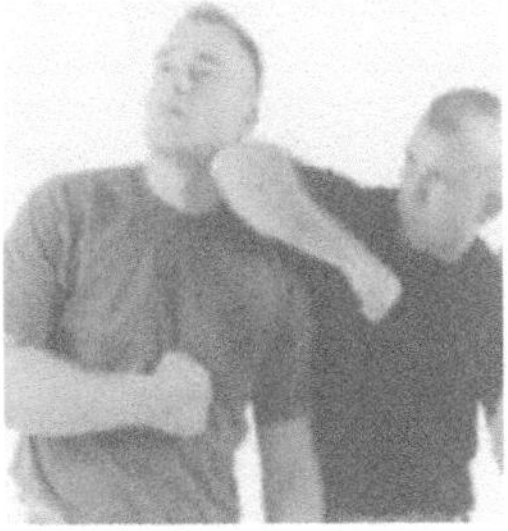

Horizontal Elbow Strike (Rear). To execute the rear horizontal elbow strike, Marines-

Tuck the right fist near the left shoulder with the palm heel facing the ground. At the same time, rotate the right shoulder forward and the left hip forward.

Thrust the right elbow horizontally rearward toward the opponent. The forearm is parallel to the ground and the hand moves toward the direction of the attack.

Rotate the right hip back and the right shoulder backward to generate additional power.

Contact the opponent with the right triceps 2 inches above the point of the elbow.

4. Strikes with the Lower Body

The legs are the body's most powerful weapons because they use the largest muscle groups to generate a strike. Legs are also less prone to injury. The feet are the preferred choice for striking because they are protected by boots. Marines use their feet, heels, and knees to execute kicks, knee strikes, and stomps.

Kicks

The purpose of a kick is to stop an opponent's attack or to create an opening in his defense in order to launch an attack. Kicks can be performed with the left (lead) leg or the right (rear) leg. Kicks with the rear leg have greater power because the hips are rotated into the attack. However, the rear leg is further away from the opponent so a strike with the rear leg will not contact the opponent as quickly as a strike with the lead leg.

Front Kick. The front kick is executed when the opponent is in front of the Marine. The front kick, delivered with either the rear or lead leg, is effective for striking below the waist. Attempting to kick higher results in diminished balance and provides the enemy with a greater opportunity to grab the leg or foot. The striking surfaces arc the toe of the boot or the bootlaces. To execute the front kick, Marines-

Raise the left knee waist high, pivot the hips into the attack, and thrust the left foot forward toward the opponent.

Contact the opponent with the toe of the left boot or bootlaces.

Return to the basic warrior stance.

Side Kick. The side kick, delivered with the lead leg, is effective for striking the knees. The side kick is executed when the opponent is to the side of the Marine. The striking surface is the outside cutting edge of the boot near the heel. To execute the side kick, Marines—

Raise the right knee waist high and rotate the right hip forward.

Thrust the right foot to the right side toward the opponent, turning the foot at a 90-degree angle to maximize the striking surface on the opponent.

Contact the opponent with the cutting edge of the right boot.

Return to the basic warrior stance.

Knee Strikes

Knee strikes are excellent weapons during the close range of close combat fighting. The knee strike is generally delivered in close.

Vertical Knee Strike. The striking surface is the thigh, slightly above the knee. To execute the vertical knee strike, Marines-

Raise the right knee and drive it up forcefully into the opponent. Power is generated by thrusting the leg upward.

Contact the opponent 2 inches above the right knee.

Horizontal Knee Strike. The horizontal knee strike is executed with the leg generally parallel with the ground while rotating the hips to generate power. It is often delivered to the peroneal nerve. The striking surface is the front of the leg, slightly above or below the knee. To execute the horizontal knee strike, Marines-

Raise the right knee, rotate the right hip forward while pivoting on the left foot, and drive the knee horizontally into the opponent.

Contact the opponent 2 inches above the right knee.

Stomps

Stomps are delivered with the feet, usually when the opponent is down. Remember, when the opponent is down, Marines take whatever target is available.

Vertical Stomp. The vertical stomp allows Marines to remain upright and balanced, to rapidly deliver multiple blows with either foot, and to quickly and accurately attack the target. It is the preferred stomp. The striking surface is the flat bottom of the boot or the cutting edge of the heel. To execute the vertical stomp, Marines—

Raise the right knee above the waist with the right leg bent at approximately a 90-degree angle.

Drive the flat bottom of the right boot or the cutting edge of the right heel down onto the opponent forcefully. At the same time, bend the left knee slightly to drop the body weight into the strike.

Ax Stomp. The striking surface of the ax stomp is the cutting edge of the heel. To execute the ax stomp, Marines-

Raise the right heel above the waist, keeping the right leg straight.

Drive the cutting edge of the right heel down onto the opponent forcefully. At the same time, bend the left knee slightly to drop the body weight into the strike.

5. Counters to Strikes

In a close combat situation, an opponent will attempt to strike Marines with punches and kicks. When an opponent uses a strike, Marines must first avoid the strike. This is accomplished by moving quickly and blocking. Next, Marines must get into an offensive position. This allows Marines to use offensive strikes to attack the opponent. Regardless of the strike, the counter to a strike requires Marines to move, block, and strike.

Move

The first step in countering a strike is to move out of the way of the strike's impact. Movement removes Marines from the opponent's intended strike point and positions Marines to attack. Movement is executed at approximately a 45-degree angle to the front or rear. Movement is always initiated from the basic warrior stance. Return to the basic warrior stance with the toe of the lead foot pointing toward the opponent once the movement is complete.

Block

Different blocks are executed based on the strike. These will be covered by individual counters.

Strike

Any of the upper body or lower body strikes can be executed as a follow-on attack or part of the counter to an opponent's strike. The follow-on strike is determined by the angle to the opponent, the position of the opponent, and the opponent's available vulnerable target areas.

Counters to Punches

Counter to a Lead Hand Punch. This counter is used when the opponent throws a lead hand punch. To execute the counter to the lead hand punch, Marines-

Step forward and to the left at approximately a 45-degree angle, moving in to the outside of the opponent's attacking arm.

Raise the left arm and block or deflect the opponent's lead hand with the palm of the hand or the meaty portion of the forearm.

Hit and stick by leaving the left arm against the opponent's right arm while stepping forward and to the left at approximately a 45-degree angle to close with the opponent.

Execute a finishing technique, such as a strike or a kick, to the opponent's exposed target areas.

Counter to a Rear Hand Punch. This counter is used when the opponent throws a rear hand punch. To execute the counter to the rear hand punch, Marines—

Step forward and to the left at approximately a 45-degree angle, moving in to the outside of the opponent's attacking arm.

Raise the left arm and block or deflect the opponent' s rear hand with the palm of the hand or the meaty portion of the forearm.

Hit and stick by leaving the left arm against the opponent's right arm while stepping forward and to the right at approximately a 45-degree angle to close with the opponent.

Execute a finishing technique, such as a strike or a kick, to the opponent's exposed target areas.

Counters to Kicks

Counter to a Front Kick (Left or Lead Leg). This counter is used when the opponent executes a front kick with his left leg. To execute the counter to a front kick, Marines-

Step forward and to the right at approximately a 45-degree angle, moving in to the outside of the opponent's striking leg.

Lower the left arm and block or deflect the opponent' s leg with the palm of the hand or the meaty portion of the forearm.

Hit and stick by leaving the left arm against the opponent's leg while stepping forward and to the left at approximately a 45-degree angle to close with the opponent.

Execute a finishing technique, such as a strike or a kick, to the opponent's exposed target areas.

Counter to a Front Kick (Right or Rear Leg). This counter is used when the opponent executes a front kick with his right leg. To execute the counter to a front kick, Marines—

Step forward and to the left at approximately a 45-degree angle, moving in to the outside of the opponent's striking leg.

Lower the left arm and block or deflect the opponent' s leg with the palm of the hand or the meaty portion of the forearm.

Hit and stick by leaving the left arm against the opponent's leg while stepping forward and to the right at approximately a 45-degree angle to close with the opponent.

CHAPTER 5

THROWS

This chapter describes all techniques for a right-handed person. However, all techniques can be executed from either side.

The Marine is depicted in camouflage utilities. The opponent is depicted without camouflage.

Marines use throwing techniques to maintain the tactical advantage and to throw the opponent to the ground during close combat. Throws apply the principles of balance, leverage, timing, and body position to upset an opponent's balance and to gain control by forcing the opponent to the ground. Throwing techniques are effective because they are size- and gender-neutral, and they rely on the momentum and power generated by the opponent rather than the strength or size of the Marine. Marines also execute a throw as a devastating attack against an opponent, possibly causing unconsciousness or broken limbs. When Marines execute throws, they must maintain balance and, simultaneously, prevent the opponent from countering a throw or escaping after being forced to the ground.

1. Turning Throw

Marines use a turning throw to take the opponent to the ground while they remain standing. A turning throw can also be executed from a stationary position. It is particularly effective if Marines and the opponent are wearing gear. To execute the turning throw, Marines-

Grasp the opponent's right wrist with the left hand.

Step forward with the right foot, place it against the outside of the opponent's right foot, and pivot so the back of the heel is next to the middle of the opponent's foot.

Hook the opponent's right arm with the right arm and pinch his arm between the biceps and forearm, touching the opponent with the body.

Pull the opponent's wrist downward, keeping it close to the body.

Pivot to the left on the ball of the foot and continue pulling downward on the opponent's wrist while rotating the wrist outward to off-balance him.

2. Hip Throw

Marines use a hip throw to take the opponent to the ground while they remain standing. A hip throw is particularly effective if the opponent is moving forward or pushing. Marines use the opponent' s forward momentum to execute the hip throw. To execute the hip throw, Marines-

Grasp the opponent's right wrist with the left hand.

Step forward with the right foot and place it against the outside of the opponent's right foot.

The back of the heel should be next to the opponent' s foot.

Step back with the left foot and rotate on the ball of the foot. The back of the heel is next to the opponent's toe. The knees are bent.

Rotate at the waist and hook the right hand around the back of the opponent's body (anywhere from his waist to his head). The side and hip should be against the opponent.

Rotate the hip against the opponent. The hips must be lower than the opponent's.

Use the right hand to pull the opponent up on the hip to maximize contact.

Pull the opponent's arm across the body and, at the same time, lift the opponent off the ground slightly by bending at the waist, straightening the legs, and rotating the body to the left.

3. Leg Sweep

Marines use a leg sweep to take the opponent to the ground while they remain standing. A leg sweep is particularly effective if the opponent is already off-balanced and moving backward or pulling on the Marine. To execute the leg sweep, Marines—

Grab the opponent's right wrist with the left hand and grab the opponent's left shoulder with the right hand.

Note: Grab the opponent's clothing or gear if his wrist and shoulder cannot be grabbed.

Step forward with the left foot and place it on the outside of the opponent's right foot.

Pull the opponent's wrist downward, close to the body, and push his shoulder backward to off-balance him.

Raise the right knee no higher than the waist.

Kick the foot past the opponent's right leg.

Use the heel of the boot to make contact with the opponent's calf (anywhere from the top of the calf down to the Achilles tendon or on the inside of the calf).

Sweep the opponent to the ground.

CHAPTER 6

CHOKES AND HOLDS

This chapter describes all techniques for a right-handed person. However, all techniques can be executed from either side.

The Marine is depicted in camouflage utilities. The opponent is depicted without camouflage.

WARNING

During training, never execute a choke at full force or full speed and never hold a choke for more than 5 seconds.

When Marines correctly perform a choke, they render an opponent unconscious in as little as 8 to 13 seconds. Chokes are easily performed regardless of size or gender. Marines must know how to apply chokes and how to counter a choke or a hold executed by an opponent.

1. Types of Chokes

There are two types of chokes: an air choke and a blood choke. An air choke closes off the airway to the lungs, thereby preventing oxygen from reaching the heart. A blood choke cuts off the blood flow to the brain. Both types can result in unconsciousness and eventual death for an opponent.

Air Choke

An air choke is performed on the opponent's windpipe (or trachea), cutting off the air to the lungs and heart. If Marines execute the air choke properly, the opponent loses consciousness within 2 to 3 minutes. Due to the length of time it takes to immobilize the opponent, air chokes are not recommended.

Blood Choke

A blood choke is performed on the opponent's carotid artery, which carries oxygen-enriched blood from the heart to the brain. The carotid artery is located on both sides of the neck. If Marines execute a blood choke properly, the opponent will lose consciousness within 8 to 13 seconds. The blood choke is the preferred choke because its intended effect (i.e., the opponent losing consciousness) can be executed quickly, ending the fight.

2. Chokes

Front Choke

Marines execute a front choke when they are facing the opponent. Marines use the opponent's lapels or collar to execute a front choke. To execute the front choke, Marines-

Grab the opponent's right lapel with the right hand, making certain that the knuckles or the back of the hand are against the opponent's carotid artery on the right side of his neck.

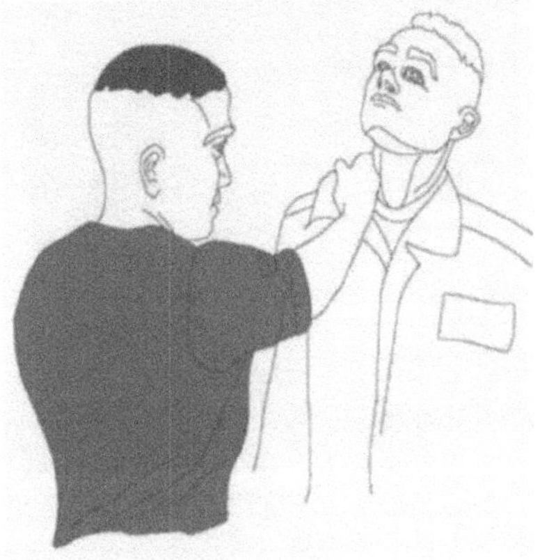

Keep the right hand against the opponent's neck, reach under the opponent's right arm with the left hand, grab the opponent's left lapel, and form an X with the wrists.

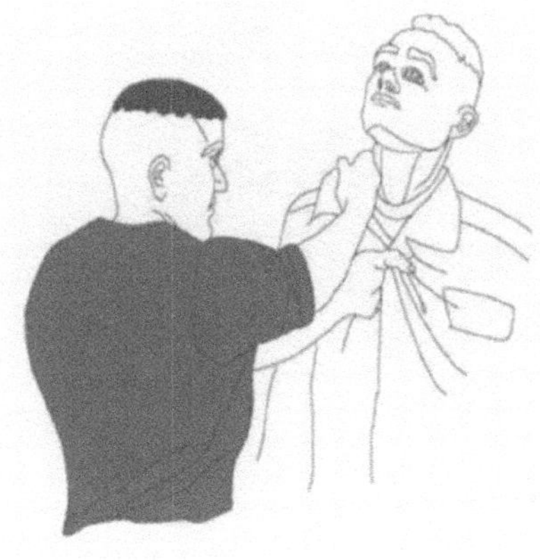

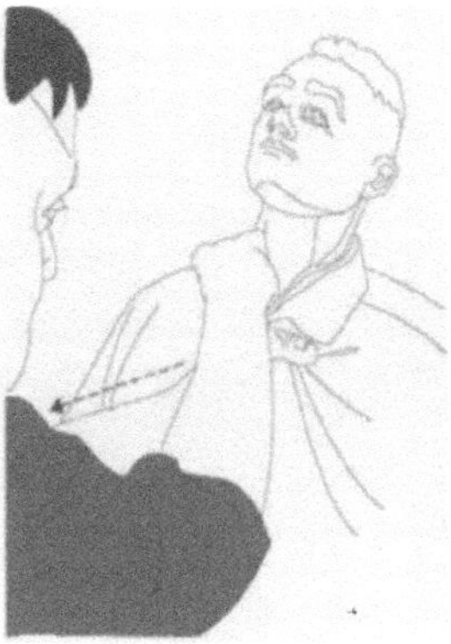

Press the right hand against the opponent's carotid artery.

Pull his lapel to the left with the left hand.

Side Choke

Marines execute a side choke when they are facing the opponent. The side choke is particularly effective when deflecting a punch thrown by an opponent. To execute the side choke, Marines—

Use the left hand to parry the opponent's arm inboard (to the inside of the opponent's reach).

Bring the right arm underneath the opponent's arm and up around the front of his neck.

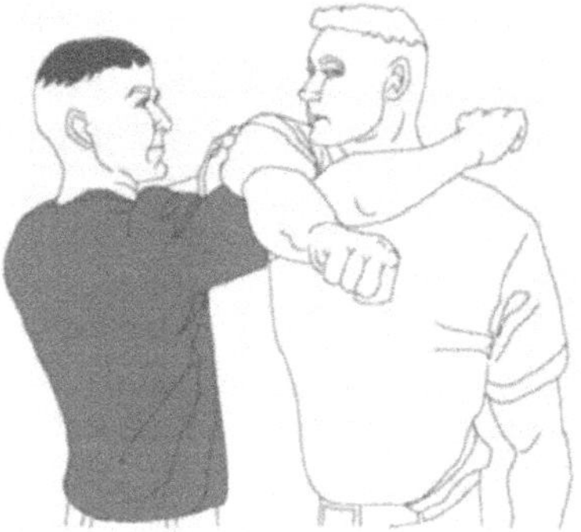

Extend the fingers, place the back of the forearm against the opponent's neck just below his ear, and press the carotid artery on his neck.

Note: If Marines are unable to place their fingers against the opponent's neck, they may use the back of the thumb or wrist.

Reach, with the left hand, around the back of the opponent's neck and clasp the hands together.

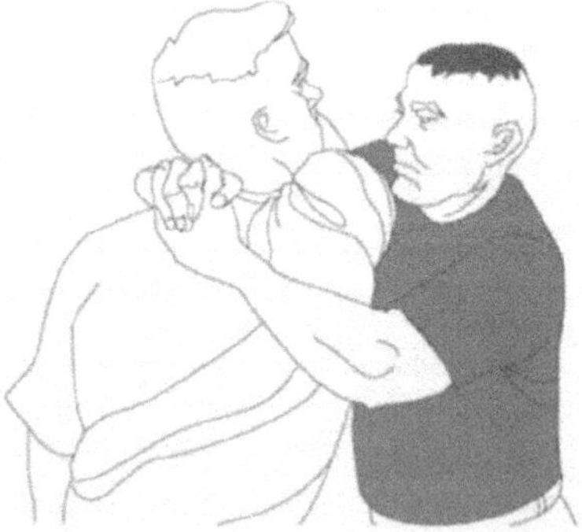

Pull the opponent toward the chest by pulling the clasped hands toward the chest.

Exert pressure on the side of the opponent's neck with the forearm.

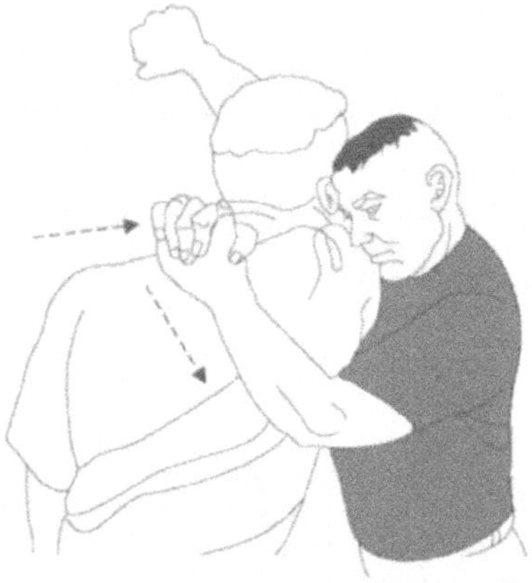

Rear Choke

Marines execute a rear choke when they are behind the opponent, the opponent is on the ground, or when they are taking the opponent to the ground. To execute the rear choke, Marines—

Reach, with the right arm, over the opponent's right shoulder and hook the bend of the arm around his neck.

Clasp both hands together.

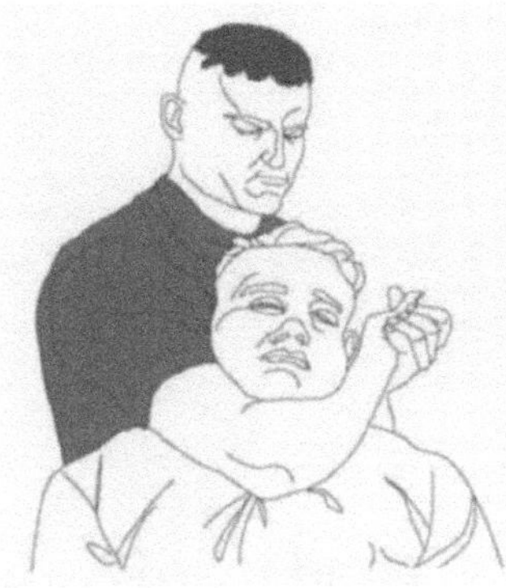

Exert pressure with the biceps and forearms on both sides of the opponent's neck on his carotid arteries.

Maintain pressure with the biceps and forearms on both sides of the neck and draw the opponent closer by drawing the right arm in.

Figure-Four Choke

A variation of the rear choke is the figure-four choke. The figure-four choke allows Marines to gain more leverage than the rear choke. If the rear choke cannot be secured, the figure-four variation is employed to increase the pressure of the choke on the opponent. To execute the figure-four variation of the rear choke, Marines—

Apply a rear choke. The Marine's body should be against the opponent's body.

Grasp the left biceps with the right hand and place the left hand against the back of the opponent's head.

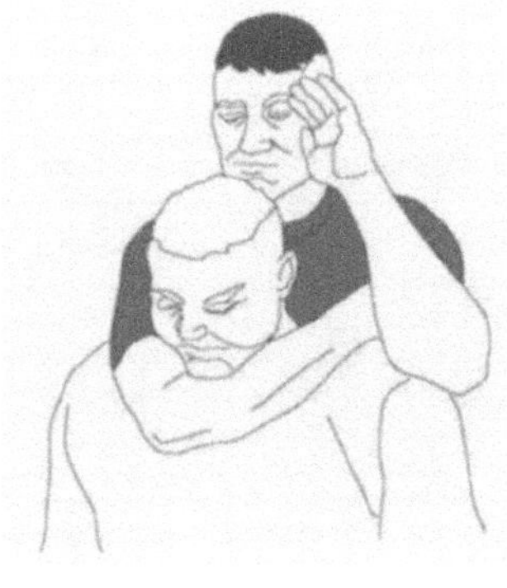

Push the opponent's head forward and down with the left hand.

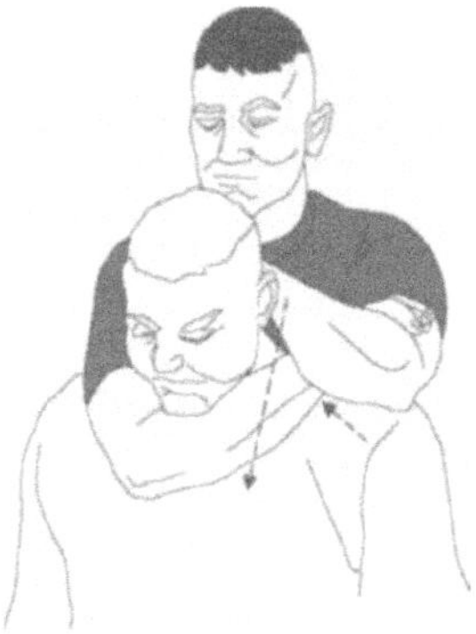

Draw the right arm in, maintaining pressure with the bicep and forearm on both sides of the opponent's neck.

3. Counters to Chokes and Holds

During a close combat situation, an opponent may apply a choke or hold on a Marine. If the opponent correctly applies a choke, a Marine quickly loses consciousness. If a choke is not executed properly, it often results in a hold, typically a bear hug or a headlock. A hold allows the opponent to control a Marine and removes the Marine's ability to attack. It is important for Marines to extract themselves from chokes and holds, regain the tactical advantage, and counter with strikes.

Although a choke causes unconsciousness in 8 to 13 seconds for a blood choke and 2 to 3 minutes for an air choke, the first movement in any counter to a choke is to clear the airway. Marines use softening techniques to loosen an opponent's grip and to clear their airway. Softening techniques are particularly effective if Marines lack the physical strength of their opponent. These techniques include groin strikes, eye gouges, foot stomps, etc. Softening techniques are not offensive; rather, they are used to loosen an opponent's hold.

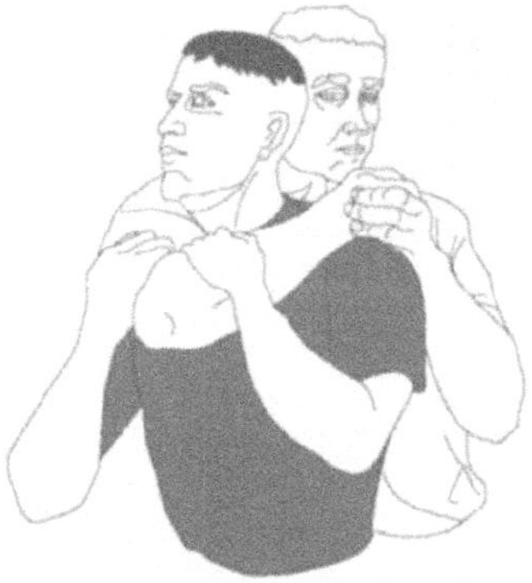

The second movement is to tuck the chin. Once the airway is clear, Marines tuck their chins to prevent the opponent from reapplying the choke.

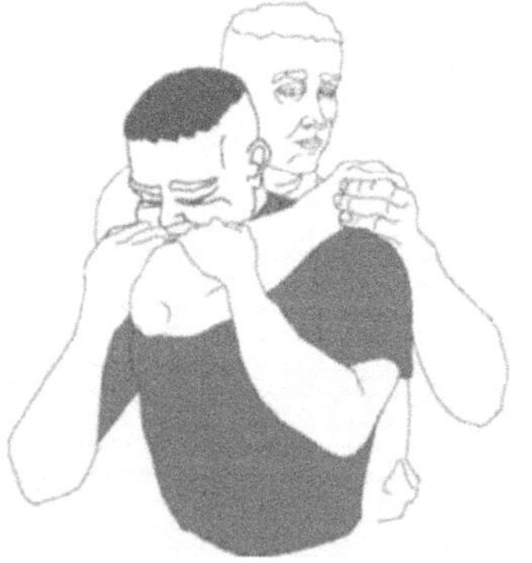

Counter to a Front Choke

Marines use a counter to a front choke when the opponent approaches from the front and uses both hands to choke a Marine around the throat. To execute the counter to the front choke, Marines—

Grasp the opponent's right forearm (where the elbow bends) with the left hand and apply downward pressure on the opponent's radial nerve with the fingers.

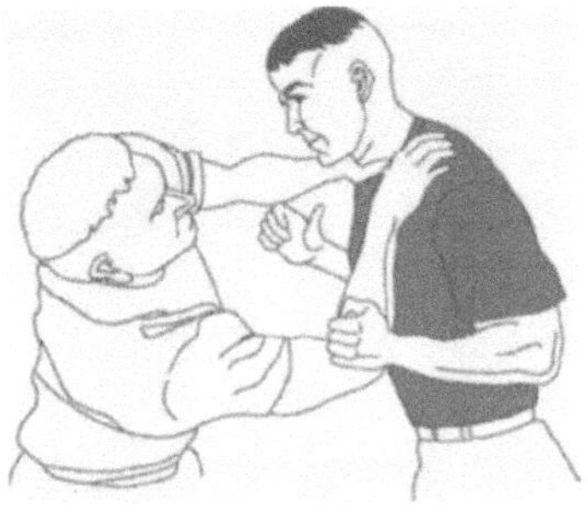

Execute a chin jab to the opponent's chin with the right hand. To generate power into the strike, bring the left foot to the outside of the opponent's right foot and rotate the hips into the strike.

Counter to a Rear Choke

Marines execute a counter to a rear choke when the opponent approaches from the rear and puts his right arm around a Marine's throat. To execute the counter to the rear choke, Marines-

Grasp the opponent's forearm (at the radial nerve) and bicep with both hands and pull down just enough to clear the airway. Once the airway is clear, tuck the chin to protect the airway and to prevent the opponent from reapplying the choke.

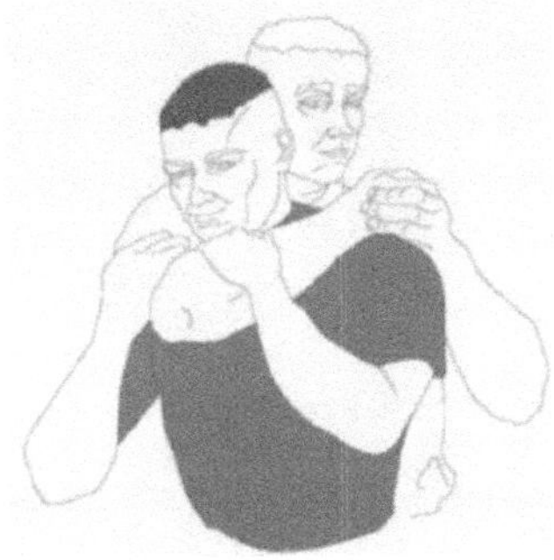

Step behind the opponent's right leg with the left foot, keeping both legs bent (almost in a squatting position).

Strike and drive the left elbow into the opponent's torso while rotating the hips and pivoting to the left, throwing the opponent back and to the ground.

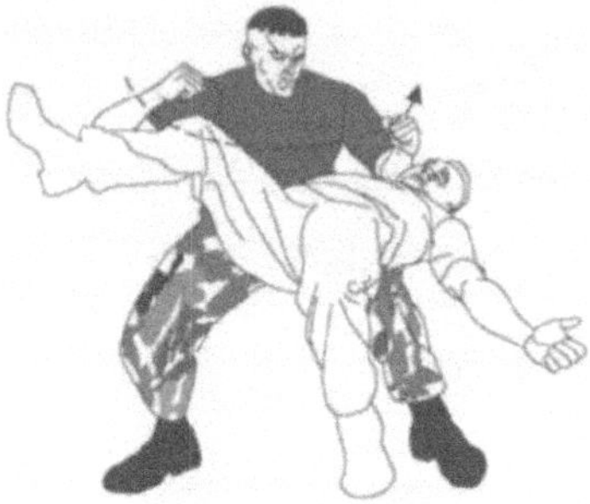

Counter to a Front Headlock

Marines use a counter to a front headlock when the opponent approaches from the front and puts his right arm around the Marine's neck, bends the Marine forward, and locks the Marine's head against his hip. To execute the counter to a front headlock, Marines—

Grasp the opponent's wrist and forearm with both hands and pull down to clear the airway. Maintain control of the opponent's wrist throughout the move. Once the airway is clear, tuck the chin to protect the airway and to prevent the opponent from reapplying the choke.

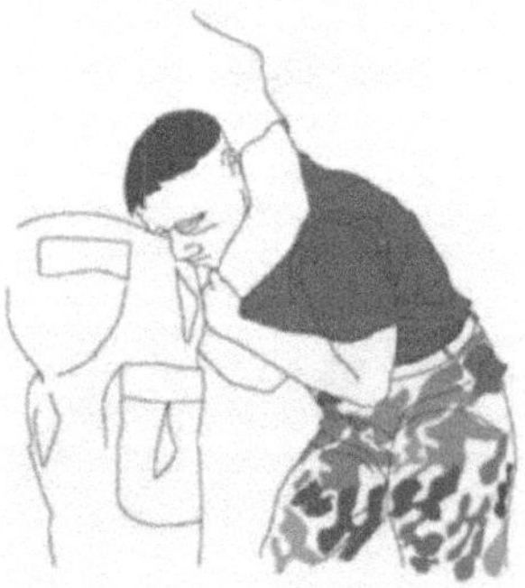

Move the right hand and arm across the opponent' s torso.

Step forward and to the left with the left foot at a 45-degree angle.

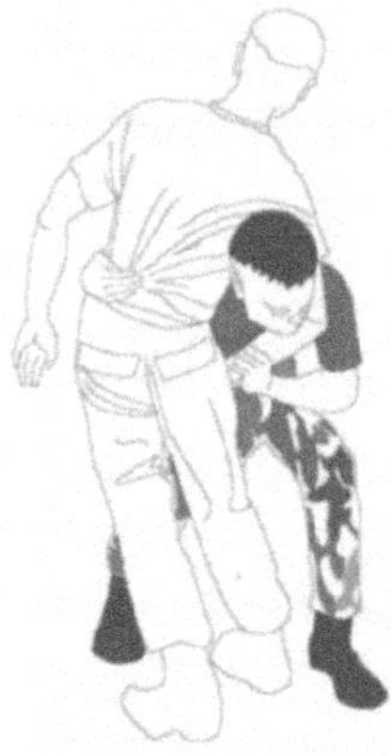

Execute a sweep with the right foot against the opponent's right leg. At the same time, push against the opponent's chest with the right arm and shoulder to generate power in the sweep.

Counter to a Rear Headlock

Marines use a counter to a rear headlock when the opponent approaches from the rear and puts his right arm around the Marine's neck, bends the Marine forward, and locks the Marine's head against his hip. To execute the counter to the rear headlock, Marines-

Grasp the opponent's wrist and forearm with the right hand and pull down to clear the airway. Once the airway is clear, tuck the chin to protect the airway and to prevent the opponent from re-applying the choke.

Reach over the opponent's right shoulder with the left arm.

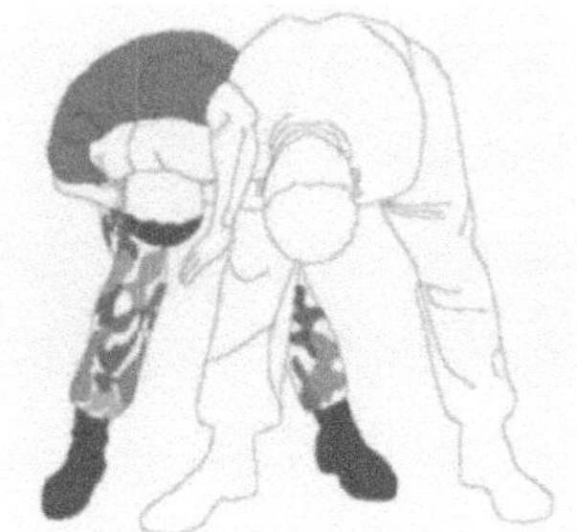

Grab any part of the opponent's face (chin, nose, eyes) and pull back while rising to a standing position.

Execute, with the right hand, a hammer fist strike to the opponent's exposed throat.

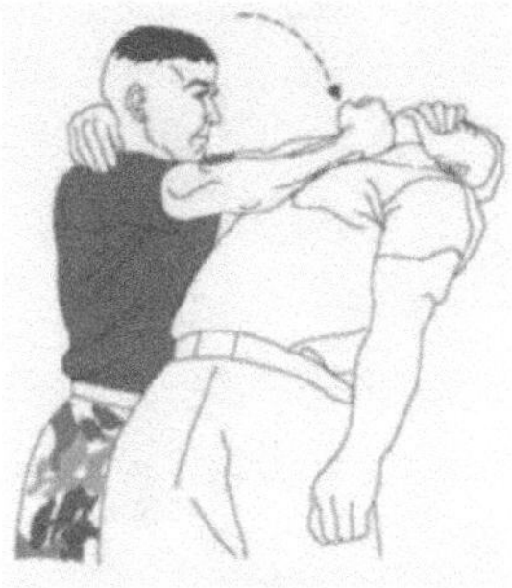

Counter to a Front Bear Hug

Marines execute a counter to a front bear hug when the opponent approaches from the front and puts both of his arms around the Marine's body, trapping the Marine's arms to the sides. To execute the counter to a front bear hug, Marines —

Step forward and to the left with the left foot at a 45-degree angle to the outside of the opponent's right leg, keeping the left leg bent.

Grasp the opponent's torso or arms to gain balance and to assist in throwing the opponent. It may be helpful to hook the opponent's right arm with the left arm.

Drive the right arm and shoulder forward and, at the same time, bring the right leg forward and sweep the opponent's right leg, bringing him to the ground.

Counter to a Rear Bear Hug

Marines use a counter to a rear bear hug when the opponent approaches from the rear and puts both of his arms around the Marine's body, trapping the Marine's arms to the sides. To execute the counter to a rear bear hug, Marines—

Step behind the opponent's right leg with the left foot, keeping both legs bent (almost in a squatting position). The left side of the body should be against the opponent's.

Pivot the hip, turning the body to the left and throwing the opponent back over the bent leg.

CHAPTER 7

GROUND FIGHTING

This chapter describes all techniques for a right-handed person. However, all techniques can be executed from either side.

In drawings, the Marine is depicted in woodland camouflage utilities; the opponent is depicted witlwut camouflage. In photographs, the Marine is depicted in woodland camouflage utilities; the opponent is depicted in desert camouflage utilities.

Marines should avoid being on the ground during a close combat situation because the battlefield may be covered with debris and there is an increased risk of injury. However, many close combat situations involve fighting on the ground. The priority in a ground fight is for Marines to get back on their feet as quickly as possible. In any ground fighting scenario, Marines will usually end up in one of four positions with the opponent. The offensive positions, in which Marines have a tactical advantage, are the guard and mount. The defensive positions, which are used as counters when the opponent has the tactical advantage, are the counter to the guard and the counter to the mount. Marines can also employ chokes during ground fighting to quickly end a fight.

1. Offensive Ground Fighting

Guard Position

Marines execute a guard position when the opponent is on top and the legs are wrapped around the opponent's legs. If the opponent is on top, he may try to choke the Marine, but the Marine still has the tactical advantage because the Marine is in a position to control the situation. To execute the guard position technique, Marines—

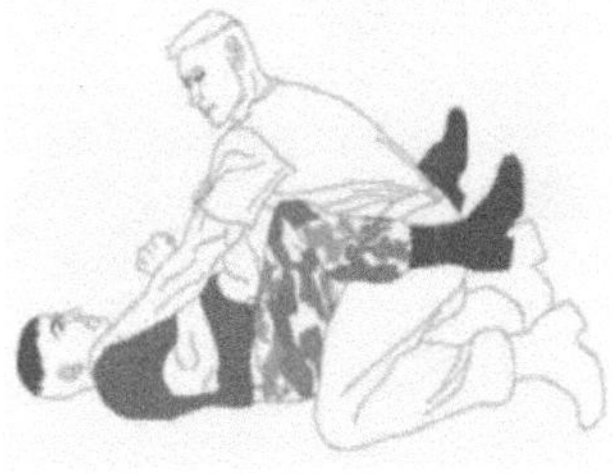

Trap the opponent's hands on the chest by crossing the hands on the chest so the forearms are resting on the opponent's forearms. Apply pressure with the elbows or forearms.

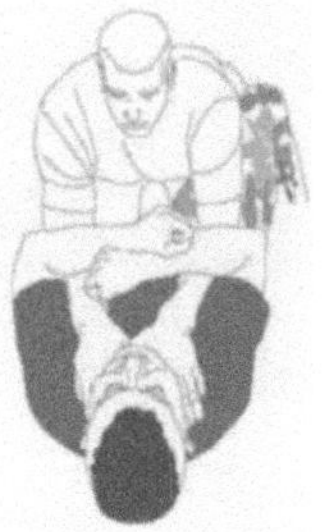

Strike the outside of the opponent's thigh with the cutting edge of the right heel. This causes the opponent to jerk to that side.

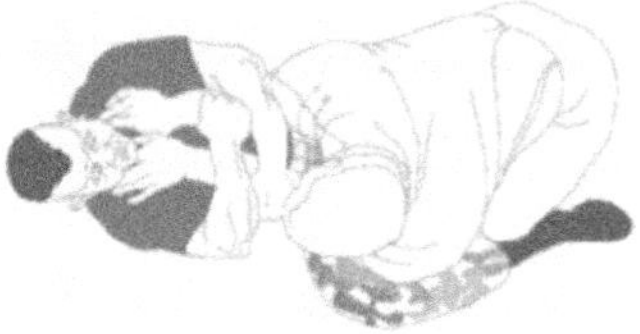

Move the head quickly to the left and swivel the hips to the right. At the same time, bring up both of the legs. Both legs are on the right side of the opponent's body.

Bring the right leg down, hooking the opponent's neck and head. Exert downward pressure to roll him over on his back. Grasp and maintain control of the opponent's left arm. Upon completion of the move, the Marine is sitting up with legs bent over the opponent while maintaining control of his arm.

Keep legs and knees bent. Maintain pressure against the opponent's neck with the back of the right foot and against his side with the left foot underneath his armpit. Squeeze the knees together, locking the opponent's arm.

Pull the opponent's arm straight up and fall back sharply, pulling his arm to the side in the direction of his little finger.

Return to standing.

Mount Position

Marines execute the mount if the opponent is lying on his back on the ground and the Marine is on top with legs wrapped around the opponent's body. This position is an offensive position because Marines are in a better position to control the opponent and to execute ground fighting techniques. To execute the mount position technique, Marines—

Grab the opponent's wrists or forearms with the hands. Hold them tightly against the torso.

Maintain control of the opponent's arms with the left arm and apply pressure to the opponent's brachial plexus (tie in) with the right hand.

Bring the left foot close to the opponent's right armpit and swing the right leg across the opponent' s head.

Keep the legs and knees bent. Maintain pressure against the opponent's neck with the back of the right foot and against his side with the left foot which is underneath his armpit. Squeeze the knees together, locking the opponent's arm.

Pull the opponent's arm straight up and fall back sharply, pulling his arm to the side in the direction of his little finger.

Return to standing.

2. Defensive Ground Fighting

Counter to the Guard

Marines use this technique if the opponent is lying on his back on the ground and the Marine is kneeling on the ground between the opponent's legs. To execute the counter to the guard position, Marines—

Strike the opponent's femoral nerve, located on the inside of the thigh, with the elbows. This forces the opponent to separate his legs.

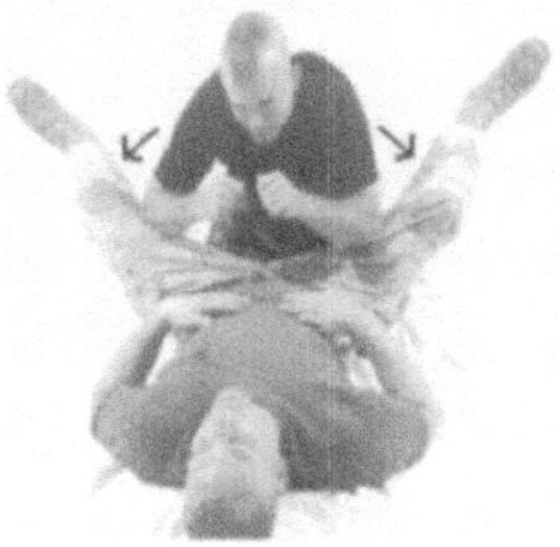

Strike the opponent's groin with the right fist.

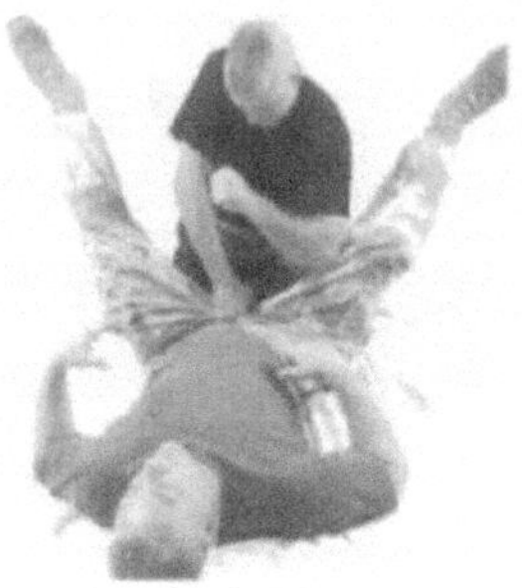

Hook the left arm underneath the opponent's right knee from the inside. Throw the opponent's leg over the head.

Duck the head quickly and move to the left.

Return to standing.

Counter to the Mount

Marines use this technique when they are lying on their back on the ground, the opponent is mounted on top of the Marine, and the opponent's legs are wrapped around the Marine. The opponent has the tactical advantage. To execute the counter to the mount position, Marines—

Grab the opponent's gear or clothing on his upper torso and pull him down close.

Use the right arm to hook the opponent's left arm, from the inside around the outside, above his elbow. With the right foot, hook the opponent's left leg or ankle.

Draw the elbow in to bend the opponent's elbow, bringing him down close. The arm must be hooked above the opponent's elbow in order to bend it.

Strike the opponent's side with the left hand.

Push the opponent over and roll him off to the right side.

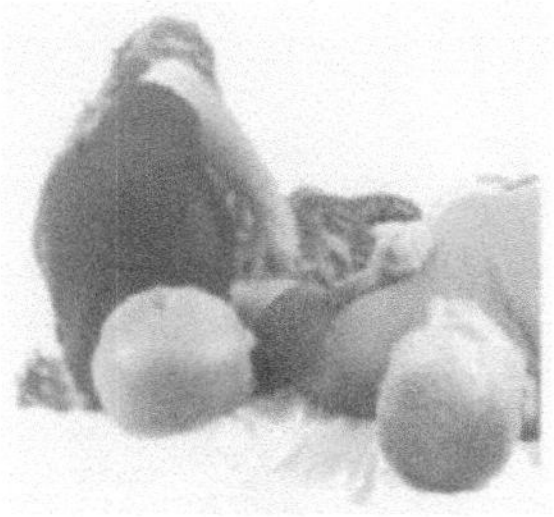

Maintain control of the opponent's hooked arm and move to a kneeling position.

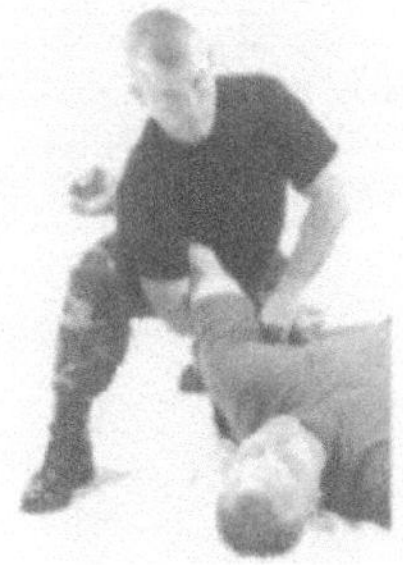

Move to a standing position while maintaining pressure on the opponent's arm and using the knee to apply pressure against the opponent's elbow.

3. Ground Fighting Chokes

WARNING

During training, never execute a choke at full force or full speed. Never hold a choke for more than 5 seconds.

The priority in ground fighting is for Marines to get back on their feet as quickly as possible. Sometimes, Marines can quickly end a ground fight by executing a choke on the opponent. When performed correctly, a choke can render an opponent unconscious in as little as 8 to 13 seconds. Chokes are easily performed regardless of size or gender. The chokes performed during ground fighting are the same as those performed while standing (see chap. 6).

Ground Fighting Front Choke

The ground fighting front choke is a blood choke performed most effectively from the mount position. The front choke employs the opponent's lapels or collar to execute the choke. To execute the ground fighting front choke, Marines—

Grab the opponent's right lapel with the right hand, making certain that the knuckles or the back of the hand are against the carotid artery on the right side of the opponent's neck.

Keep the right hand pressed against the opponent's carotid artery, reach under the right arm with the left hand, and grab the opponent's left lapel, forming an X with the wrists.

Keep the right hand pressed against the opponent's carotid artery and pull the opponent's left lapel to the left with the left hand.

Ground Fighting Side Choke

The ground fighting side choke is a blood choke performed from the mount position. The ground fighting side choke is particularly effective when the opponent raises his arms and places them on the Marine's chest or throat. To execute the ground fighting side choke, Marines-

Use the left hand to parry the opponent's right arm inboard (to the inside of the opponent's reach). Bring the right arm underneath the opponent's arm and up around the front of his neck.

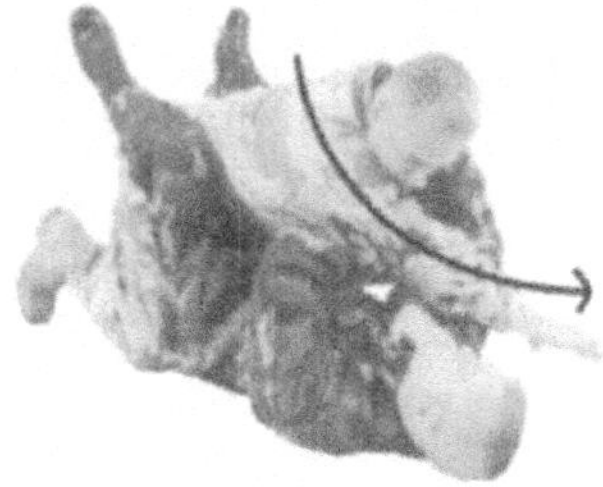

Keep the fingers extended, place the back of the forearm against the opponent's neck just below his ear and press the carotid artery on his neck. Depending on the Marine's or the opponent's body configuration, the back of the thumb or wrist may press against the opponent's neck.

Use the left hand to reach around the back of the opponent's neck and clasp the hands together.

Pull the opponent toward the chest. Use the fore-arm to exert pressure on the side of his neck. This is done by pulling the clasped hands toward the chest.

Ground Fighting Rear Choke

The ground fighting rear choke is a blood choke performed when Marines are behind the opponent. To execute the ground fighting rear choke, Marines—

Wrap the legs around the opponent with the left foot against the inside of his left thigh and the right foot against the inside of his right thigh.

Use the right arm to reach over the opponent's right shoulder and hook the bend of the arm around his neck.

Use the left hand to clasp both hands together.

Exert pressure with the biceps and forearms on both sides of the opponent's neck. This applies pressure to the opponent's carotid arteries. While maintaining pressure with the biceps and fore-arms, draw the opponent closer by drawing the right arm in. To increase the effectiveness of the choke, lean backward by arching the back and pulling the opponent back. At the same time, push the feet against the opponent's thighs.

Figure-Four Choke

A variation of the ground fighting rear choke is the figure-four choke. The figure-four choke allows Marines to gain more leverage on the rear choke. If the rear choke cannot be secured, the figure-four variation may be applied to increase the pressure of the choke on the opponent. To execute the figure-four variation of the rear choke, Marines—

Apply the ground fighting rear choke. The chest should be against the opponent's back.

Grasp the left biceps with the right hand and place the left hand against the back of the opponent's head.

Use the left hand to push the opponent's head forward and down.

Draw the right arm in while maintaining pressure with the biceps and forearm on both sides of the opponent's neck.

Increase the effectiveness of the choke by arching backward and pushing the feet against the opponent's thighs. Continue to pull the opponent back with the right arm while exerting pressure forward with the left hand against the opponent's head.

CHAPTER 8

NONLETHAL TECHNIQUES

This chapter describes all techniques for a riglit-handed person. However, all techniques can be executed from either side.

In drawings, the Marine is depicted in woodland camouflage utilities; the opponent is depicted without camouflage. In photographs, the Marine is depicted in woodland camouflage utilities; the opponent is depicted in desert camouflage utilities.

The Marine Corps' involvement in military operations other than war—e.g., humanitarian, peacekeeping, or evacuation missions—has greatly increased. These missions require skills that span the spectrum of conflict and support operations within a continuum of force. But the Corps' day-to-day existence also demands a responsible use of force. Nonlethal techniques are among the skills Marines use to apply a responsible use of force.

1. Unarmed Restraints and Manipulation

Marines operate within a continuum of force, particularly in support of peacekeeping or humanitarian types of missions. In these situations, Marines must act responsibly to handle a situation without resorting to deadly force. Unarmed restraints and manipulation techniques including joint manipulation, come-alongs, and takedowns can be used to control a subject without resorting to deadly force. Marines must train to become proficient in nonlethal techniques and to respond in a responsible manner. These techniques are referred to as compliance techniques, and they are applied in the third level in the continuum of force.

WARNING

During training, never apply the techniques for unarmed restraints and manipulation at full force or full speed. Use a slow and steady pressure to avoid injury.

Compliance Techniques

Compliance techniques are unarmed restraint and manipulation techniques used to physically force a subject or opponent to comply. Compliance can be achieved through the close combat techniques of-

Pain compliance using joint manipulation and pressure points. (Pain compliance is the initiation of pain to get compliance on the part of the subject.)
Come-along holds.

Principles of Joint Manipulation

Joint manipulation is used to initiate pain compliance and gain control of a subject. It involves the application of pressure on thejoints (elbow, wrist, shoulder, knee, ankle, and fingers). Pressure is applied in two ways:

In the direction in which the joint will not bend. For example, joints such as the knees and elbows only bend in one direction and when pressure is applied in the opposite direction, pain compliance can be achieved. Beyond the point where the joint stops naturally in its range of movement (i.e., it no longer bends).

Since each joint has a breaking point, Marines should apply slow steady pressure only until pain compliance is reached. Continued pressure will break the joint and may escalate the violence of the situation.

Joint manipulation also uses the principle of off-balancing. A subject can be better controlled when he is knocked off balance.

Wristlocks

A wristlock is a joint manipulation that can be applied in a number of ways to achieve pain compliance. The wrist rotates in a number of directions and will bend in a single direction until its movement stops naturally. In a wristlock, pressure is exerted beyond that point by bending or twisting the joint. A wristlock is executed when an opponent tries to grab Marines or is successful in grabbing Marines or their equipment. A wristlock can also be performed by Marines if they wish to initiate control of an opponent.

Basic Wristlock. A basic wristlock is executed when Marines grab the opponent's left hand with the right hand. To execute the basic wristlock, Marines—

Use the right hand to grab the opponent's left hand by placing the thumb on the back of the opponent's hand so that the Marine's knuckles are facing to the left.

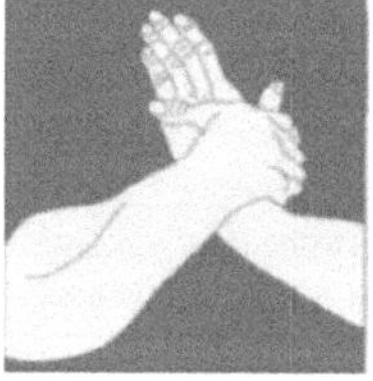

Hook the fingers across the fleshy part of the opponent's palm below the thumb. The fingers are used to anchor the hand so leverage can be applied to twist and bend the joint.

Exert downward pressure with the thumb to bend the opponent's joint. Rotate the opponent's hand to the right to twist the joint.

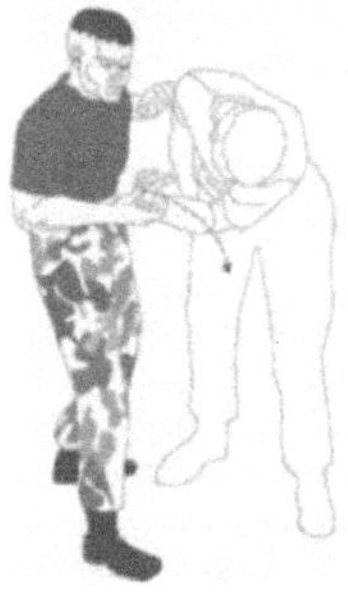

Use the left hand to further control the opponent.

Step in to the opponent to keep the opponent's hand in close to the body to control him and provide more leverage on the wristlock.

Note: When executing the basic wristlock with the left hand, the Marine grabs the opponent's hand so that the Marine's knuckles are facing to the right, and then rotates and twists the opponent's hand to the left.

Reverse Wristlock. A reverse wristlock is executed when Marines grab the opponent's right hand with the right hand. To execute the reverse wristlock, Marines—

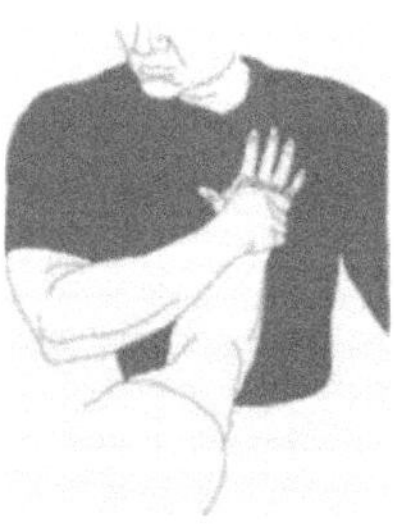

Place the right palm on the back of the opponent's right hand and wrap the fingers across the fleshy part of his palm below his little finger.

Twist the opponent's hand to the right while stepping in to place his hand against the chest. Apply downward pressure on the opponent's hand against the chest. Leave the opponent's hand on the chest to fully control the subject and to gain leverage.

Lean forward to use body weight to add additional pressure to the joint.

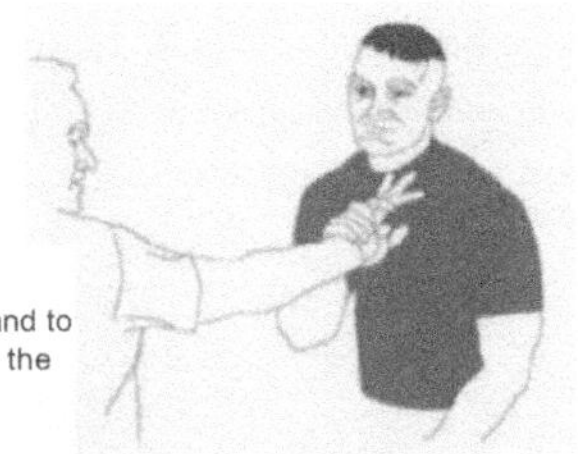

Use the left hand to further control the opponent.

Two-Handed Wristlock. Both hands can be used in the wristlock to maximize the leverage and pressure needed to bend and twist the joint. To execute the two-handed wristlock, Marines—

Place both thumbs on the back of the opponent's hand, thumbs crossed.

Hook the fingers of both hands around the fleshy part of the opponent's palm on both sides of his hand.

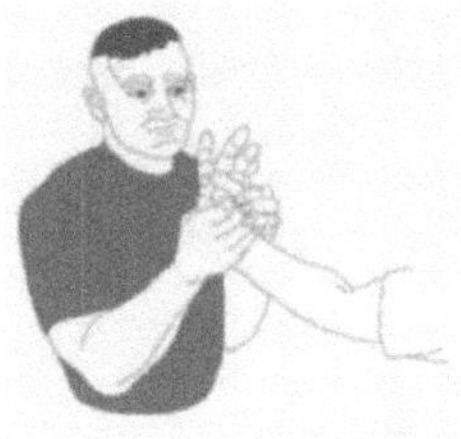

Step into the opponent and apply pressure downward on the back of his hand to bend the joint and rotate his wrist away from the body to twist the joint.

Enhanced Pain Compliance on Wristlock. Enhanced pain compliance techniques are applied in the third and fourth levels in the continuum of force. Additional pain be applied to a wristlock by-

Adding downward pressure to the elbow with the other hand or elbow by using the fingers to pull in on the opponent's radial nerve located on the inside of the forearm. When pressure is added to the opponent's radial nerve, his direction can be controlled.

Applying pressure against the opponent's finger joint to bend it in a direction it cannot bend (i.e., splitting the fingers).

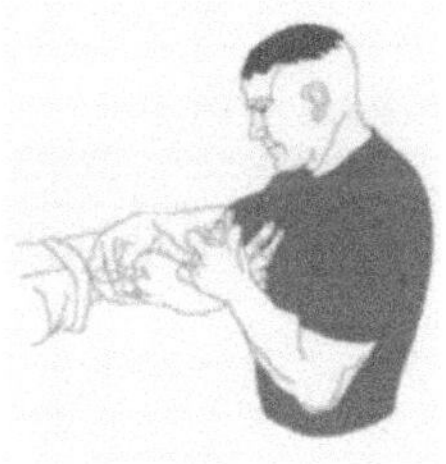

Come-Along Holds

Marines use a come-along hold to control and move an opponent.

Escort Position. A common come-along hold is the escort position. To execute the escort position, Marines—

Face the opponent. Use the left foot to step forward at a 45-degree angle. Turn to face the right side of the opponent.

Use the right hand to firmly grasp the opponent's right wrist. With the left hand, firmly grasp the opponent's right triceps.

Position the opponent's controlled arm diagonally across the torso, keeping his wrist against the right hip. The Marine should be standing to the right of and behind the opponent.

Note: This technique works well when escorting an opponent on either the right or left side. Take caution when escorting an opponent by ensuring his controlled hand is not in a position to grab the holstered weapon. The preferred escort position is from the Marine's left side, so that the opponent is kept further away from the weapon.

Wristlock Come-Along. To execute the wristlock come-along, Marines-

Use the left hand to execute a basic wristlock. Incorporate the right hand in a two-handed wristlock for more control.

Maintain pressure on the opponent's wrist with the right hand, step forward, and pivot around to stand next to the opponent.

Release the left hand, quickly reach under the opponent's arm from behind, and grab his hand.

Use the left hand and apply downward pressure on the opponent's wrist.

Controlling Technique. The following controlling technique is used when an opponent grabs the Marine's wrist. To execute the technique, Marines—

Trap the opponent's hand with the palm of the other hand.

Rotate the opponent's trapped hand up and on his forearm while maintaining downward pressure on his trapped hand.

Apply downward pressure with both hands until the opponent is taken to the ground.

Armbars

An armbar is a joint manipulation in which pressure is applied on a locked elbow, just above the joint, in the direction the joint will not bend. An armbar has to be locked in quickly, but still requires a slow, steady pressure to gain compliance.

Basic Armbar. To execute a basic armbar, Marines-

Use the right hand to grab the opponent's right wrist.

Bring the left hand down on or above the opponent's elbow joint. To gain additional leverage, pivot to face the opponent.

Use the left hand to apply downward pressure on or above the opponent's elbow joint while pulling up on his wrist.

Armbar from a Wristlock. To execute an armbar from a wristlock, Marines-

Use the right hand to grab the opponent's right hand and execute a reverse wristlock.

Bring the left hand down on or above the opponent's elbow joint.

Use the left hand to apply downward pressure on or above the opponent's elbow joint while pulling up on his wrist.

Takedowns

A takedown is used to bring an opponent to the ground to further control him.

Takedown From a Wristlock Come-Along. To take the opponent to the ground from a wristlock come-along, Marines—

Use the right foot to push down on the opponent's calf or Achilles tendon.

Maintain control of the opponent's wrist and elbow and apply a slow, steady pressure to bring him to the ground.

Armbar to a Takedown. This technique is used to take a noncompliant opponent to the ground from an armbar. To execute the armbar takedown, Marines—

Use the right hand to execute a reverse wristlock.

Bring the left hand or forearm down on or above the opponent's elbow joint.

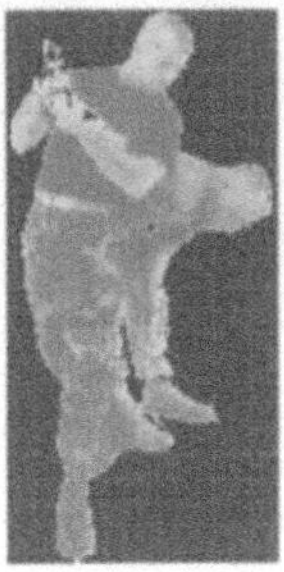

Pivot so the back is facing the opponent and, at the same time, lift the left elbow and slide the body so it is against the opponent, placing the armpit high above the opponent's elbow joint.

Lean back, placing the body weight on the opponent's arm until he complies or is taken to the ground.

Note: This technique may break the opponent's arm. Therefore, this technique should not be employed if the objective is a nonlethal takedown.

Wristlock Takedown. This technique is used to take a noncompliant opponent to the ground from a basic wristlock and to put him in a position where he can be handcuffed, if necessary. To execute the takedown, Marines-

Use the right hand to execute a basic wristlock.

Incorporate the left hand in a two-handed wristlock.

Apply downward pressure on the wristlock, pivot on the ball of the right foot, and quickly turn to the right to take the opponent to the ground.

Continue to apply pressure on the wristjoint as the opponent lands on his back with his arm straight in the air.

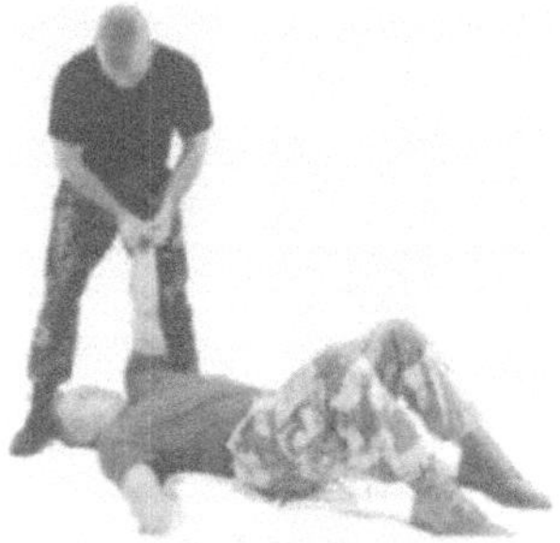

Slide the left foot under the opponent's back, underneath his armpit.

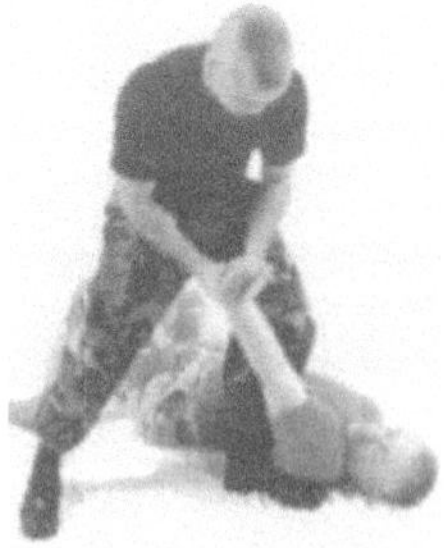

Apply pressure with the knee against the opponent's triceps while pulling back on his arm.

Continue applying pressure with the knee on the opponent's arm. Pivot and step around the opponent's arm to roll him on his stomach.

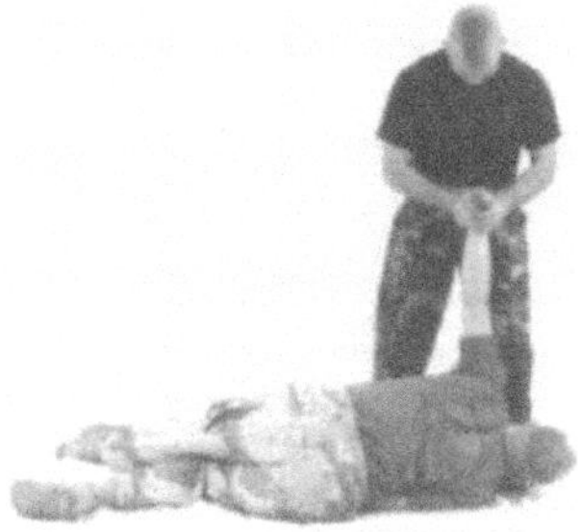

Kneel down with one knee on the opponent's back. The other knee is placed on the opponent's neck and shoulder on either side of his arm. Apply inward pressure with the knees to lock his arm in place.

Tell the opponent to put his other hand in the middle of his back. Bring the opponent's controlled hand to the center of his back.

Escort Position Takedown. This technique can be used to control a noncompliant opponent from the escort position. To execute the escort position takedown, Marines—

Lock the opponent's arm straight across the body while rotating his wrist away from the body.

Use the left hand or forearm and apply downward pressure above the opponent's elbow where the triceps meet.

Step back with the right foot and, keeping the opponent's hand controlled against the hip, pivot to the right while continuing to apply downward pressure on his arm to bring him to the ground.

Note: This technique works well when escorting a subject on the Marine's right or left side. When taking down a subject from the right side, step back and pivot to the left.

2. Nonlethal Baton

A baton or nightstick can be an effective compliance tool when used correctly. Batons can be used defensively (blocking), offensively (striking), and as a restraining device when needed. In the fourth level of the continuum of force (assaultive [bodily harm]), defensive tactics include baton or nightstick blocks and blows. Blows to the head or other bony parts of the body are considered deadly force. When deadly force is not authorized, Marines must be able to employ blocks, strikes, and restraints with the baton with the minimum of force.

Grips

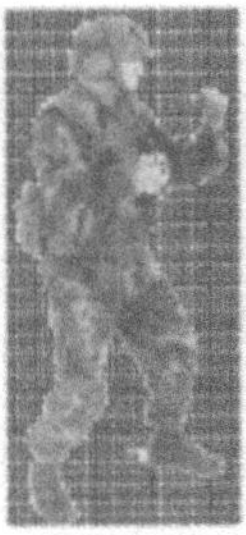

One-Handed Grip. To execute the one-handed grip, Marines—

Use the right hand to grasp the lower end of the baton, about 2 inches from the end.

Wrap the thumb and index finger around the baton so they are touching one another. The grip on the baton should be firm, but natural.

Two-Handed Grip. To execute the two-handed grip, Marines-

Use the right hand to grasp the lower end of the baton, about 2 inches from the end. Wrap the thumb and index finger around the baton so they touch one another.

Use the left hand to grasp the upper end of the baton, palm down, about 2 inches from the end. The hands should be approximately 10 to 12 inches apart.

Stance and Method of Carry

The basic warrior stance serves as the foundation for initiating nonlethal baton techniques. The method of cany provides effective defensive positions with a wide range of options to control a combative opponent.

One-Handed Carry. To execute the one-handed carry, Marines—

Grip the baton using the one-handed grip.

Elevate the baton, with the gripping hand at a level between the belt and shoulder.

Keep the left hand in the position of the basic warrior stance.

Two-Handed Carry. This carry is effective for blocks. To execute the two-handed carry, Marines-

Grip the baton using the two-handed grip.

Elevate the baton, with the left hand higher than the right hand.

Orient the weapon toward the subject.

Movement

In a nonlethal confrontation, movement may be made to create distance between Marines and the opponent or to close the gap to control him. When facing an opponent, Marines move in a 45-degree angle to either side of the opponent. Moving at a 45-degree angle is the best way to both avoid an opponent's strike and to put Marines in the best position to control the opponent.

Blocking Techniques

One-handed blocks are used when carrying the baton in the one-handed carry. The same one-handed blocks used in combative stick techniques (see page 3-6) apply in nonlethal baton engagements. The blocks include blocks for a vertical strike, a forward strike, and a reverse strike. Because the baton is often carried with two hands, there are also two-handed blocks that are effectively used from this carry. Two-handed blocks are discussed in the following subparagraphs.

High Block. Marines execute a high block to deter a downward vertical attack directed at the head and shoulders. To execute the high block, Marines—

Raise the baton up to a level even with or above the head. The baton should be in a horizontal position to block the blow. The fingers of the left hand should be open and behind the baton.

Place the baton perpendicular to the opponent's striking surface to absorb the impact of the blow.

Bend the elbows slightly to help absorb the impact of the blow. The arms should give with the strike of the blow.

Low Block. Marines execute a low block to deter an upward vertical attack directed at the abdomen, groin, or torso. The opponent's blow can be delivered by a foot, knee, or fist. To execute the low block, Marines—

Lower the baton to a level even with or below the groin. The baton should be in a horizontal position to block the blow. The fingers of the left hand should be open and behind the baton.

Place the baton perpendicular to the opponent's striking surface to absorb the impact of the blow.

Bend the elbows slightly to help absorb the impact of the blow. The arms should give with the strike of the blow.

Right Block. Marines execute a right block to deter an attack directed at the head, neck, flank, or hip. The opponent's blow can be delivered by a foot, knee, fist, or elbow. To execute the right block, Marines-

Thrust the baton in a vertical position to the right side.

Pivot to the right by stepping with the left foot and pivoting off the ball of the right foot. Rotate the hips and shoulders into the direction of the block. The fingers of the left hand should be open and behind the baton.

Place the baton perpendicular to the opponent's striking surface to absorb the impact of the blow.

Bend the elbows slightly to help absorb the impact of the blow. The arms should give with the strike of the blow.

Left Block. Marines execute a left block to deter an attack directed at the head, neck, flank, or hip. The opponent's blow can be delivered by a foot, knee, fist, or elbow. To execute the left block, Marines—

Thrust the baton in a vertical position to the left side.

Pivot to the left by stepping with the right foot and pivoting off the ball of the left foot. Rotate the hips and shoulders into the direction of the block. The fingers of the left hand should be open and behind the baton.

Place the baton perpendicular to the opponent's striking surface to absorb the impact of the blow.

Bend the elbows slightly to help absorb the impact of the blow. The arms should give with the strike of the blow.

Middle Block. Marines execute either a left or right block to deter an attack directed at the face, throat, chest, or abdomen. To execute the middle block, Marines—

Thrust the baton in a vertical position straight out in front of the body. The fingers of the left hand should be open and behind the baton.

Place the baton perpendicular to the opponent's striking surface to absorb the impact of the blow. The baton should be held with the left hand forward of the right.

Bend the elbows slightly to help absorb the impact of the blow. The arms should give with the strike of the blow.

Restraining Technique

The strong-side armlock is used to restrain an opponent who is not compliant. To execute the strong-side armlock, Marines—

Use the right hand to run the baton up under the opponent's left armpit, parallel to the ground.

Use the right foot to step forward at a 45-degree angle to the left side of the opponent. The baton is across his forearm.

Use the right hand to drive the baton forward and up so the action bends the opponent's arm behind his back. At the same time, continue moving around him to get behind him.

Place the baton on the opponent's forearm with the thumb and/or fingers and apply pressure to his forearm. At the same time, grasp the other end of the baton with the left hand.

Pull up on the low end of the baton with the right hand. At the same time, push down on the top end of the baton with the left forearm, reaching around with the left hand to grasp the opponent's biceps or shoulder.

Continue exerting downward pressure with the left forearm while pulling back on the opponent's biceps with the left hand. This places the opponent in a position where he is controlled and can be moved.

Apply pressure with the foot against the bend in the opponent's leg above his calf. This lowers the opponent to the ground rather than throwing him to the ground and risking severe injury.

Striking Target Areas

Marines must avoid striking an opponent in the head, neck, or other bony parts with the baton because this is considered deadly force and can lead to serious bodily injury or death. Instead, the legs, arms, and buttocks are target areas that are considered nonlethal.

Legs. Primary targets are the thighs and lower legs. Avoid striking the knee and ankle joints because this can cause permanent damage.

Arms. Primary targets are the upper arms. Avoid striking the shoulder and elbow and wrist joints because this can cause permanent damage.

Buttocks. Primary targets are the buttocks. Avoid striking any other part of the torso, including the chest, rib cage, spine, tail bone, and groin because strikes to these areas can cause permanent damage or death.

One-Handed Striking Techniques

One-Handed Forward Strike. A forward strike follows either a horizontal line or a downward diagonal line using a forehand stroke. To execute the one-handed forward strike, Marines-

Stand facing the opponent with the baton carried in a one-handed cany.

Place the right hand palm up, swing the baton from right to left, and make contact with the opponent.

One-Handed Reverse Strike. A reverse strike follows either a horizontal line or a downward diagonal line using a backhand stroke. To execute the one-handed reverse strike, Marines—

Stand facing the opponent with the baton carried in a one-handed carry.

Bend the right arm and cross the arm to the left side of the body. The baton should be close to or over the left shoulder.

Place the right hand palm down, swing the baton from left to right, and make contact with the opponent.

Two-Handed Striking Techniques

Two-Handed Forward Strike. This strike is an effective follow-up to a middle block or left block. To execute the two-handed forward strike, Marines—

Pull back with the left hand while driving the right hand forward toward the opponent. The baton should be horizontal to the ground. Power is generated by stepping forward with the right foot and rotating the right hip and shoulder into the strike.

Contact the opponent with the end of the baton.

Two-Handed Reverse Strike. This strike is an effective follow-up to a middle block or left block. To execute the two-handed reverse strike, Marines—

Pull back with the right hand while driving the left hand forward toward the opponent. The baton should be horizontal to the ground. Power is generated by stepping forward slightly with the left foot and rotating the left hip and shoulder into the strike.

Contact the subject with the end of the baton.

Front Jab. This strike is effective for countering a frontal attack. It can also be executed as a quick poke to keep a subject away. To execute the front jab, Marines thrust both hands forward in a quick jab. The baton is held either horizontal to the ground or at a slight downward angle.

Rear Jab. This strike is effective for countering a bear hug from the rear. To execute the rear jab, Marines thrust both hands rearward in a quick jab. The baton is held either horizontal to the ground or at a slight downward angle.

APPENDIX A

PUGIL STICK TRAINING

A pugil stick is a training device used to simulate a rifle bayonet so that effective, but safe, training can be conducted to build proficiency of rifle bayonet techniques. Pugil stick training builds on the techniques used to throw punches. Pugil stick training is the only "full contact" training provided to Marines in the Close Combat Program. Pugil stick training teaches Marines to function when faced with stress and violence, and it prepares them to deliver a blow and take a blow. It also provides them with the physical and mental skills vital to success on the battlefield.

1. Pugil Stick Training

Design

A pugil stick consists of a stick wrapped in padding at both ends that can be gripped like a rifle. The pugil stick is approximately the same weight and length of an unloaded rifle with a bayonet attached.

Holding the Pugil Stick

The pugil stick is held in the same manner as the sen ice rifle. All movements come from the basic warrior stance. To hold the pugil stick correctly, Marines—

Use the right hand to grasp the lower end of the pugil stick overhanded.
Use the left hand to grasp the upper end of the pugil stick underhanded.
Use the right forearm to lock the lower end of the pugil stick against the hip.
Orient the blade end of the pugil stick toward the opponent.

Safety Equipment

The following safety equipment must be worn during any pugil stick bout.

Groin Protection. Groin protection protects the groin from an accidental blow. It should be pulled high around the waist to protect the groin area, with the concave portion against the body.

Flak Jacket. The flak jacket provides protection to the body. It is worn completely fastened.

Neck Roll. The neck roll prevents whiplash if Marines receive a blow to the head. The neck roll further supports the head and protects the neck from blows. The neck roll is worn above the flak jacket and below the helmet. The tied end of the neck roll faces the front.

Helmet. A regulation football helmet protects the face and head. The helmet must fit snugly and the chin straps must be adjusted and snapped.

Mouthpiece. The mouthpiece is worn on the upper teeth to protect the teeth.

Gloves. Marines wear gloves to protect the hands if pugil sticks do not have gloves built-in.

Pugil Stick Screening

Prior to pugil stick training, instructors must ask participating Marines the following questions. If a Marine answers "yes" to any question, he must see the corpsman or a competent medical authority who evaluates him and determines whether or not the he can participate in the training.

Are you on light duty?
Are you restricted to running shoes by a corpsman or a doctor?
Have you fought in a pugil stick bout within the last 7 days?
Have you received a blow to the head within the last 7 days?
Have you had a concussion within the last 6 months?
Have you had dental surgery within the last 24 hours?
Do you have stitches or staples on your body?
Have you had a shoulder or head injury within the last 5 years?
Are you taking a prescription drug?
Do you have an ear infection or current sinus infection?
Have you had a broken bone within the last 6 months?

Safety Personnel

The following safety personnel are required to conduct pugil stick training:

One close combat instructor must officiate the bout.
One close combat instructor trainer, commissioned officer, or staff noncommissioned officer must be in the training area to serve as range safety officer.
One corpsman must be in the training area.

Training Area

To prevent injury, Marines train on areas with soft footing (i.e., sand or grass). Training mats are not recommended because feet can stick to the mats, prohibiting movement or causing joint injuries by twisting a knee. Bouts should not take place on a hard surface area; e.g., a flight deck or parking lot. A boxing ring may be used to conduct pugil stick bouts; ring dimensions can vary as long as there is ample room to execute the techniques in the training area.

Second Impact Syndrome

Second Impact Syndrome occurs when a second blow to the head produces a second concussion that occurs within 1 week following a previous concussion (before recovery from the first concussion). Second Impact Syndrome causes rapid brain swelling and can cause death. Therefore, there must be 7 days between pugil stick bouts to reduce the risk of severe injury resulting from Second Impact Syndrome. The 7 day separation between pugil stick bouts significantly reduces the possibility of injury, particularly in someone who may have suffered a brain injury or concussion but shows no symptoms.

Any Marine who experiences headaches or the following symptoms after training must be examined by appropriate medical personnel:

Blurred vision.
Ringing in the ears.
Dilation of the pupils.
Slurred speech.
Bleeding from ears or mouth.
Swelling in head or neck area.
Any unnatural discoloration in head or neck area.

The Marine should not be allowed to participate in pugil stick training or any other activity where a heavy blow might be sustained for a minimum of 7 days after the headache or other symptoms have subsided.

Safety Measures

The following safety measures must be followed:

Gear must be worn properly throughout training.
Contact lenses or glasses will not be worn.
False teeth will be removed from the mouth.
Nothing will be worn around the neck except the neck roll.

Competition among groups of Marines is authorized as long as it does not overshadow training objectives or compromise safety procedures.

Safety and proper techniques are paramount.

Safety is more important than competition.

2. General Rules and Regulations Governing Pugil Stick Bouts

Instructors and Support Personnel Requirements

At a minimum, one close combat instructor observes a pugil stick bout. For safety purposes, it is better to have two instructors judging a bout because each instructor can fully observe each of the fighters. The best position for observation is to the right of a fighter. This allows the instructor to see the fighter's facial expression and body movement. The instructor's position must not interfere with the fight.

In addition, one commissioned or staff noncommissioned officer (to serve as range safety officer) and one corpsman will be in the training area.

Prior to the Bout

Prior to the bout, Marines—

Are paired according to height, weight, and gender.

Wear the proper safety equipment.

Once Marines have properly donned the safety equipment, they wait for instructions from the close combat instructor. On the command of the close combat instructor, a pair of Marines will enter the ring. Once in the ring, the close combat instructor inspects each Marine for the proper safety equipment.

Note: There should be enough gear so that when two Marines are training, two other Marines can be donning safety equipment for the next bout.

During the Bout

The bout begins when the close combat instructor blows the whistle. All strikes are directed above the waist. Upon hearing a whistle blast, all fighting immediately ceases.

Stopping a Bout

There are two reasons for stopping a bout: delivery of a scoring blow or an unsafe condition. The close combat instructor trainer, close combat instructor, or the range safety officer may stop the bout at any time an unsafe condition is observed.

Scoring Blow. A scoring blow is an offensive technique delivered to a vulnerable area of an opponent with sufficient force and precision to be considered as a disabling or killing blow. Scoring blows are not judged solely on the degree of force with which a blow is delivered, but on the accuracy and techniques employed. A scoring blow is defined as—

A straight thrust with the blade end of the weapon (red end of the pugil stick) to the opponent's face mask or throat.

A slash to the side of the opponent's helmet (below the ear) or neck with the red end of the pugil stick.

A heavy blow to the opponent's head with an authorized technique (i.e., buttstroke, smash) using the butt (black end) of the pugil stick.

When a scoring blow is delivered, the close combat instructor blows the whistle to stop the bout.

Unsafe Condition. The bout will be stopped as soon as an unsafe condition exists. An unsafe condition exists when a Marine is unable to defend himself, loses his balance and falls down on one or two knees or falls down completely, shows instability (e.g., buckling at the knees), loses muscular tension in his neck and his head snaps back or to one side, or appears disoriented. An unsafe condition also exists if a Marine lets go of one end of the pugil stick, the equipment (e.g., helmet, neck roll) falls off, or a Marine fails to use the proper techniques.

If any of these conditions occur—

The close combat instructor trainer, close combat instructor, range safety officer, or anyone supervising training stops the bout and separates the two Marines.
The corpsman evaluates the possibly injured Marine to determine if the fight can continue. The corpsman—

- Ensures that he is alert and responsive, both verbally and physically.
- Talks to him to see if he is coherent.
- Makes certain he comprehends and replies to verbal questions.
- Makes certain his speech is not slurred.
- Checks his physical signs.
- Ensures his eyes are focused and not dazed or glazed.
- Ensures his legs are not wobbly or shaky.

The pugil stick is not used as a baseball bat. The use of unauthorized techniques will result in expulsion from the bout. Only techniques taught in bayonet training are authorized. These techniques include-

Straight thrust.
Buttstroke (horizontal and vertical).
Smash.
Slash.
Parry.
Blocks (low, high, left, right).

After the Bout

A third whistle blast is used prior to identifying the Marine who delivered the killing blow.

A fourth whistle blast is used prior to demonstrating the killing technique used by the Marine who won the bout.

3. Directions for Making Pugil Sticks

Close combat instructor trainers are taught to make pugil sticks to standard specifications at school. It is recommended that all pugil sticks be made under the supervision of a close combat instructor trainer. Until the Marine Corps acquires a standardized pugil stick that units can order through the supply system, instructors use the following information and instructions to make pugil sticks locally.

Supplies

The following supplies are needed to make one pugil stick:

One 45" by 2" circular, oak, wood dowel.
Four 9" wide by 27" long foam pad pieces to make two inner handguards of double thickness.
Four 11" wide by 19" long foam pad pieces to make two outer handguards of double thickness.
Canvas material to cover inner and outer handguards.
One foam pad cut 1/4" thick by 7" wide by 13" long to make the centerguard.
One 8" wide by 14"long piece of canvas material to make a pugil stick sock.
Canvas material to make two pugil stick end caps. Ideally, two different color fabrics should be used to simulate the bayonet and buttstock ends of a rifle. Common colors for end caps are black and red. Each end cap is a different color to help the officiating certified close combat instructor determine the scoring blow during the pugil stick bout.
Two 1/4" thick by 11" wide by 54" long foam pad pieces to make two ends.
Two foam pads cut 1/4" thick by 3" wide by 14 1/2" long to make two end plug inserts.
One can of 3M Photo MountQ spray adhesive (NSN 8040-01-120-4009)to adhere foam padding to the stick. (Spray adhesive is available from self-service.)
One roll of duct tape or riggers tape (NSN 7510-00-266-5016) to protect foam pad and

canvas from wear and to reinforce gluing action of spray adhesive. (Duct tape or riggers tape is available from self-service.)
One roll para-cord or 5-50 cord to tie both the inner and outer handguards together.
One roll of thread strong enough to sew canvas material together.
Sixteen grommets, one for each of the four corners of the inner and outer handguards.
One foam sleeping mat (NSN 8465-01-109¬3369). Use unserviceable foam sleeping mats obtained from Defense Rcutilization and Marketing Office.

Tools

The follow ing tools are needed to make a pugil stick:

Tape measure, ruler, or yardstick.
Scissors.
Grommet machine.
Knife.
Sewing machine.

Directions

To make a pugil stick, Marines perform the following steps.

Preparing Materials

Step 1. Cut foam sleeping mat into the dimensions listed above. Cut mats to double thickness.

Step 2. Use spray adhesive to adhere the two 9" by 27" pieces of foam together. Use the spray adhesive to adhere the two 11" by 19" pieces of foam together. Let adhesive dry before using the double-thick foam for handguards.

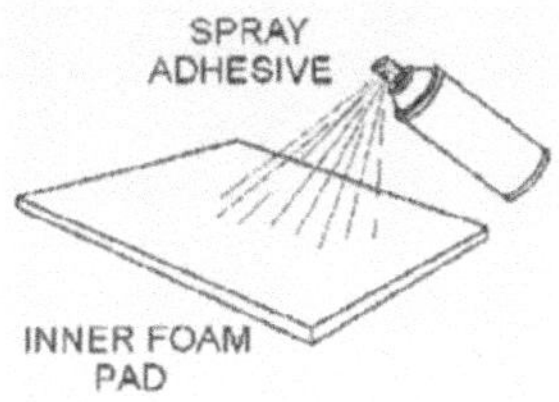

Step 3. Cut canvas material to make two coverings for the inner handguards. Cut material to make two coverings for the outer handguards. Two inner handguards and two outer handguards are required for each pugil stick.

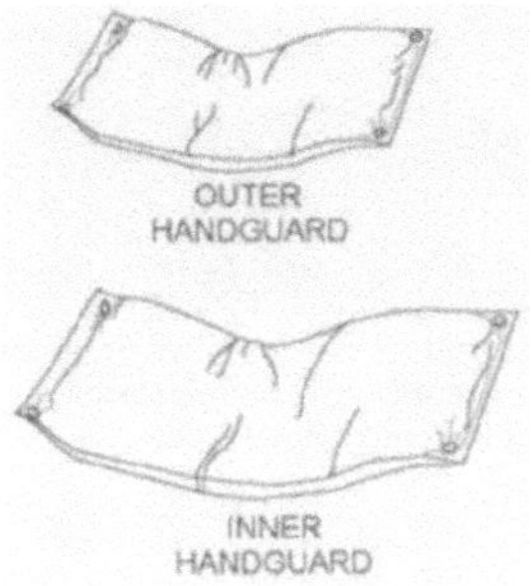

Step 4. Use a sewing machine to make a sleeve out of the 9" by 27" canvas. Sew down three sides. Insert foam pad inside the canvas sleeve and then sew down the fourth side. Use the grommet machine to create holes reinforced with metal grommets in each of the four corners. Once finished, the inner handguards arc complete.

Note: It is recommended that the Fabric Repair Shop be used to sew canvas sleeves and apply grommets.

Step 5. Repeat step 4 to make the other inner handguard and both outer handguards.

Constructing Ends, End Plugs, and End Caps

Step 1. Apply spray adhesive on approximately 8 inches of one end of the dowel. This adhesive serves to glue the foam pad to the wooden dowel.

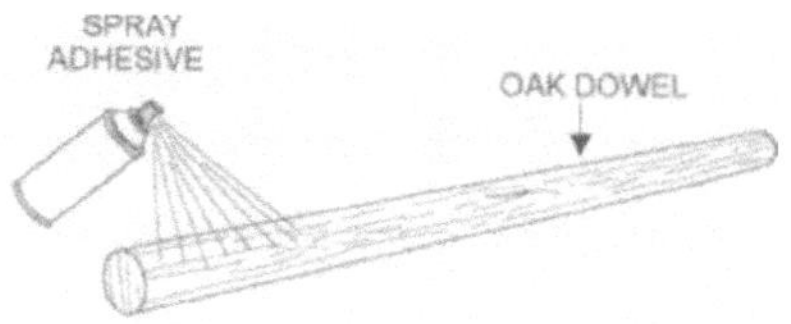

Step 2. Apply spray adhesive on one side of the 11" by 54" foam pad. This causes the foam pad to stick to itself when it is rolled around the end of the wooden dowel. Ensure 3 inches of the rolled foam pad extends past the wooden dowel for stability.

Step 3. Apply spray adhesive on one side of 3" by 14 1/2" foam pad. Roll pad into a tight cylinder.

Step 4. Apply spray adhesive on the exposed end of the dowel and fill the hole in the end with the end plug. The end plug prevents the wooden dowel from protruding beyond the edge of the rolled foam padding. This is an important safety precaution to prevent injury.

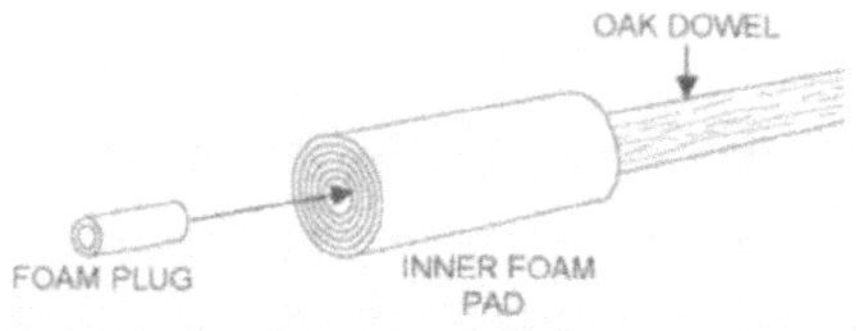

Step 5. Use duct tape or riggers tape to secure and reinforce the foam pad once the end plug is inserted in the end.

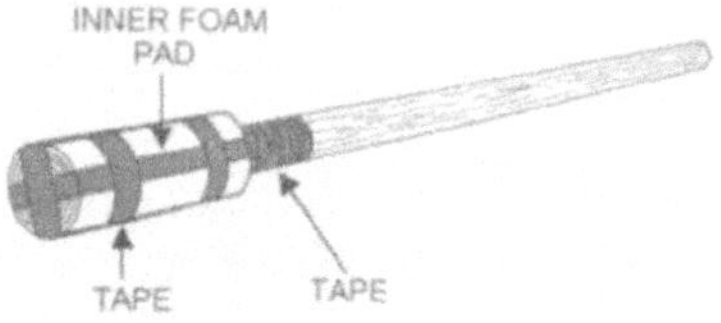

Step 6. Measure the dimensions of the ends (typically, the length of the cylinder should be 12 inches long to allow for taping). Cut and sew canvas material in the shape of a cylinder with one open end. Once complete, this becomes the end cap. The end cap is a colored sleeve that covers the rolled and taped foam pad at each end of the pugil stick.

Note: It is recommended that the Fabric Repair Shop be used to make the end cap.

Step 7. Slide one of the end caps over the rolled foam pad and tape it to the dowel.

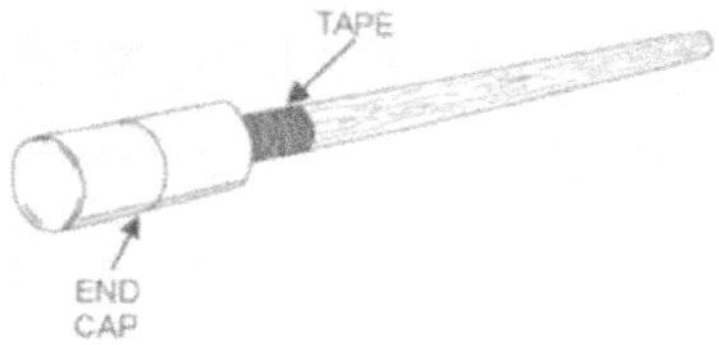

Building the Centerguard

Step 1. Spray one side of the 7" by 13" foam pad with spray adhesive and wrap it tightly around the center of the dowel.

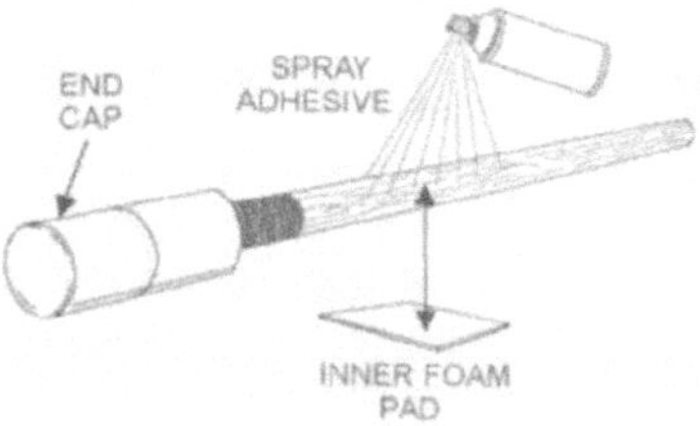

Step 2. Wrap duct tape or riggers tape around the foam pad to secure.

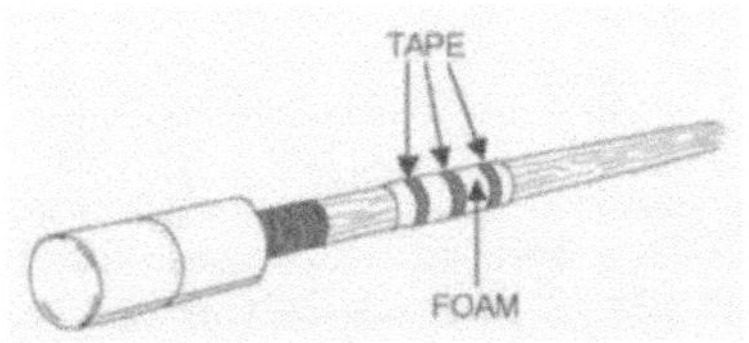

Step 3. Use a sewing machine to make a tubular pugil stick sock out of the 8" by 14" piece of canvas. Slide the pugil stick sock over the rolled foam pad and secure it with the duct tape or riggers tape. Duct tape or riggers tape protects the edges of the centerguard and reinforces the gluing effects of the spray adhesive.

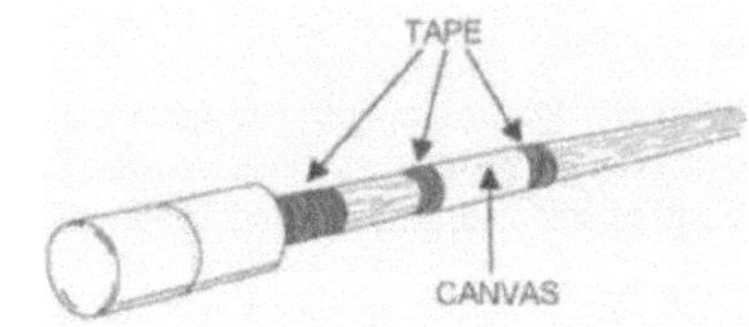

Note: It is recommended that the Fabric Repair Shop be used to make the pugil stick sock.

Finishing the Opposite End Cap

After one end cap is complete and the centerguard is finished, then repeat steps 1 through 7 in Constructing Ends, End Plugs, and End Caps for the other end of the pugil stick. Ensure that the opposite color end cap is used.

Fastening Handguards

Step 1. Place the inner handguards (longer piece) around the dowel. The inner handguard is placed on first because it protects the wrist. One inner handguard is needed for the left hand and one for the right hand.

Step 2. Fold the inner handguard around the pugil stick and run the para-cord or 5-50 cord through the metal grommets.

Step 3. Tighten down the corner ends of the inner handguards by pulling on the para-cord or 5-50 cord.

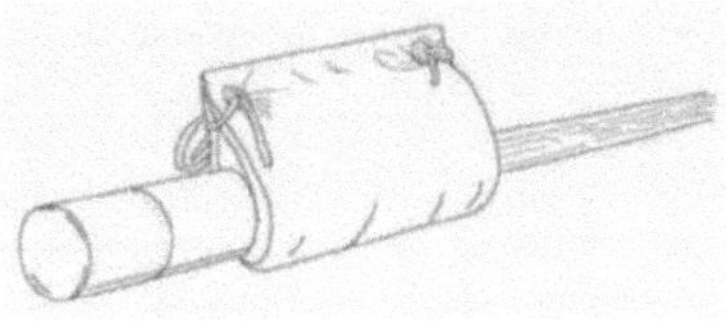

Step 4. Tie off the cord with a square knot above the metal grommets.

Step 5. Place the outer handguards (shorter piece) over the inner handguards (longer piece). Repeat steps 1-4 for the other handguard. The outer handguards provide additional protection to the knuckles and fingers and provides additional stability for the inner handguards.

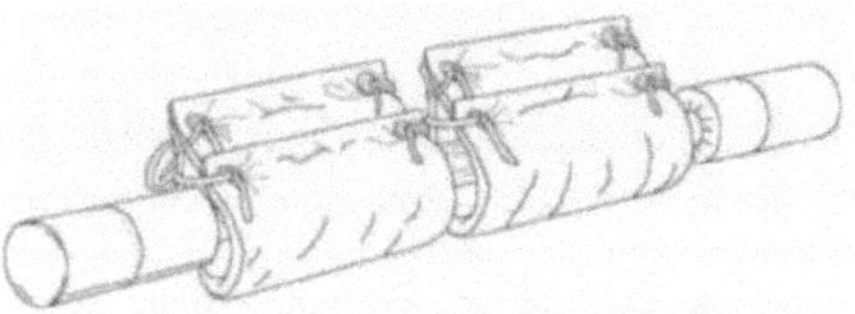

Serviceability Inspection

Prior to any training, all pugil sticks are inspected to ensure they are fit for use. If any of the following conditions exist, the pugil stick is unsafe and will not be used:

Stick protrudes through the end caps.
Frayed or worn edges exist on canvas surfaces.
The stick can be felt through the pads or the padding is too soft.
End caps or handguards are not securely attached.

APPENDIX B

SAFETY PRECAUTIONS DURING TRAINING

1. General Safety Precautions

When training close combat techniques, certain safety precautions must be adhered to to prevent injuries.

Most training should be conducted on a training area with soft footing such as a sandy or grassy area. If training mats are available, they should be used. A hard surface area is not appropriate for close combat training.

All techniques should be executed slowly at first. Marines can increase the speed of execution as they become more proficient. Marine-on-Marine training that requires contact (chokes, throws, ground fighting, and unarmed restraints and manipulation) should not be executed at full force or full speed.

If a technique is applied to the point that a Marine is uncomfortable, the Marine must "tap out." This indicates immediate release of the pressure being applied or to immediately stop the technique. The Marine "taps out" by firmly tapping his hand several times on any part of the opponent's body that will get his attention or by saying *stop*.

Second Impact Syndrome occurs when a second concussion develops within hours, days, or weeks following a prior concussion (and before recovery from the first concussion). Second Impact Syndrome causes rapid brain swelling and can cause death. Marines who experience headaches or other symptoms following training must be examined by appropriate medical personnel. These symptoms can include, but are not limited to, blurred vision, ringing in the ears, dialation of the pupils, bleeding from the ears or mouth, slurred speech, swelling in head or neck area, or any unnatural discoloration of head or neck. They should not be allowed to participate in pugil stick training or any other activity where a heavy blow might be sustained for a minimum of 7 days after the symptoms have subsided.

2. Safety Precautions for Individual Techniques

Falls

When training Marines to fall, they should progress from the ground, to a kneeling or squatting position, and then to a standing position. This ensures they are comfortable and understand the technique before progressing to executing falls from a higher profile. This instructional technique vastly reduces the risk of injury.

Strikes and Punches

When training Marines to strike and punch, they begin by executing the techniques "in the air." As they become more proficient, they execute strikes on equipment (when available) such as an air shield or a heavy bag. At no time should they be permitted to execute strikes on another student.

Chokes and Ground Fighting Chokes

When training Marines to execute chokes, they will not apply pressure to the opponent's throat during training because the trachea and windpipe can be crushed. During training, Marines should practice the proper procedures for blood chokes. They should never execute air chokes. At no time during training should a choke be applied at full force or full speed. No choke should be held for longer than 5 seconds.

Unarmed Restraints and Manipulation

When training Marines to execute unarmed restraints and manipulation techniques, they utilize slow and steady pressure. Never apply these techniques at full force or at full speed.

Pugil Stick Training

A blow to the head during training will bruise the brain. A second blow to the head can cause death. This is known as the Second Impact Syndrome. There must be 7 days between pugil stick training to prevent injury or death.

TABLE OF CONTENTS

Chapter 1. The M9 Service Pistol

Chapter 2. Weapons Handling

Chapter 3. Fundamentals of Pistol Marksmanship

Chapter 4. Pistol Firing Positions and Grip

Chapter 5. Use of Cover and Concealment

CHAPTER 1
THE M9 SERVICE PISTOL

The Marine Corps' uses the M9 service pistol as a defensive weapon. It is a semiautomatic, magazine-fed, recoil-operated, double-action pistol. The M9 service pistol's magazine holds 15 rounds and is chambered with a 9-mm ball (NATO M882 round). The M9 service pistol has a maximum effective range of 50 meters (54.7 yards).

Note: The procedures in this chapter are written for right-handed Marines; left-handed Marines reverse instructions as necessary.

Functional Capabilities

When the M9 service pistol's safety lever is moved to the firing position, it can be fired in either a single-action or double-action mode. It is designed to fire one round each time the trigger is pulled. When the last round is fired, the slide automatically locks to the rear.

Single-Action Mode

The single-action mode allows the pistol to be fired when the hammer is cocked; single action requires the hammer to be cocked to the rear before the trigger is pulled. The hammer can be manually cocked or mechanically cocked. The hammer is mechanically cocked after the first shot is fired. See figure 1-1.

Figure 1-1. Single-Action Mode.

Double-Action Mode

The double-action mode causes the hammer to move to the rear as the trigger is being pulled. See figure 1-2.

Figure 1-2. Double-Action Mode.

Nomenclature

See figure 1-3 on page 1-2.

Major Components

The M9's major components consist of the slide assembly, barrel assembly, and receiver. See figure 1-4 on page 1-3.

Slide Assembly

The slide assembly houses the firing pin, firing pin block, striker, extractor, and loaded chamber indicator, and it cocks the hammer during recoil.

Extractor

The extractor pulls the brass from the chamber after the round is fired.

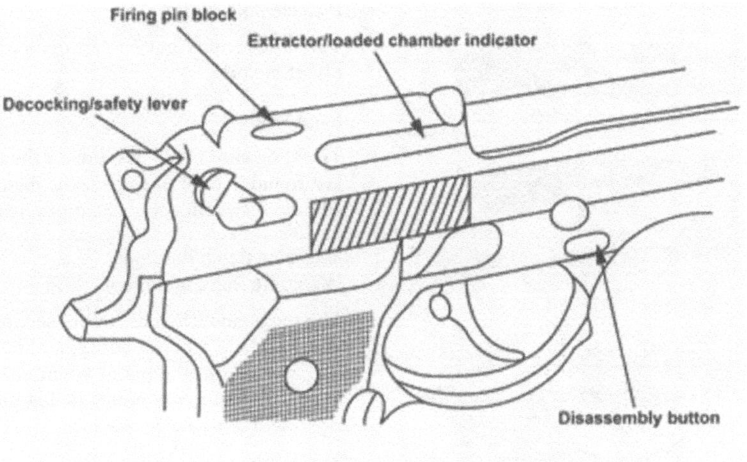

Right Side View

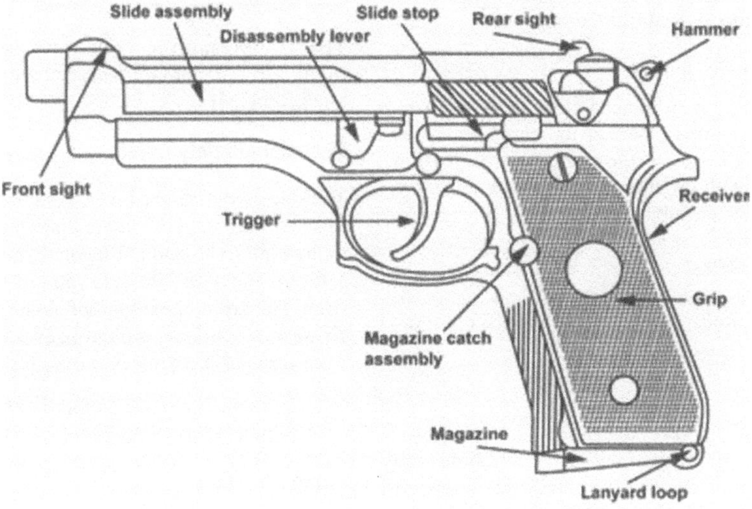

Left Side View

Figure 1-3. M9 Service Pistol.

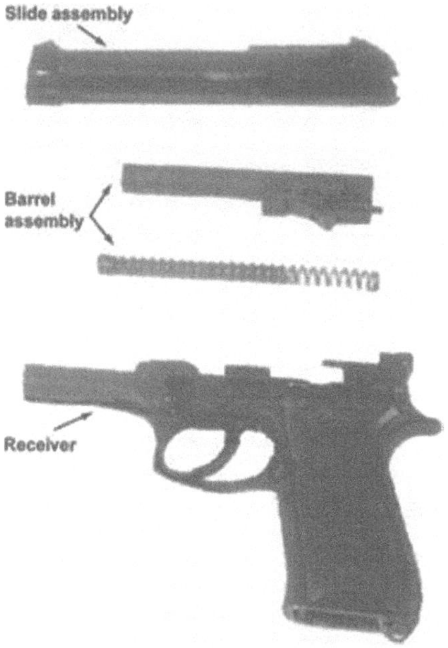

Figure 1-4. Major Components.

Loaded Chamber Indicator

When a round is in the chamber, the upper surface of the loaded chamber indicator protrudes from the right side of the slide. This protrusion can be felt with the finger, verifying that there is a round in the chamber.

Barrel Assembly

The barrel assembly houses the round for firing, directs the projectile, and locks the barrel in position during firing.

Receiver

The receiver supports the major components, controls the functioning of the pistol, and holds the magazine in place. The front and back straps of the grip are grooved vertically to ensure that the hand does not slip while firing. The receiver consists of the disassembly button, slide stop, and magazine catch assembly.

Disassembly Button

The disassembly button permits quick disassembly of the pistol.

Slide Stop

The slide stop holds the slide to the rear after the last round is fired. It can also be manually operated to lock the slide to the rear or release the slide.

Magazine Catch Assembly (Magazine Release Button)

The magazine catch assembly secures the magazine in place when loading, and it releases the magazine from the pistol when unloading. The magazine catch assembly is designed for both right- and left-handed Marines.

> *Note:* Reversal of the magazine catch assembly for left-handed Marines can be performed by a qualified armorer.

Safety Features

The safety features of the M9 service pistol include the decocking/safety lever, firing pin block, and half-cock notch.

Decocking/Safety Lever

The decocking/safety lever, commonly referred to as the safety, permits safe operation of the pistol by both right- and left-handed Marines. As the safety is moved to the safe (down) position, the firing pin striker moves out of alignment with the firing pin. This movement prevents the pistol from firing as the hammer moves forward.

> *Note:* In the fire (up) position, a red dot is visible, indicating that the pistol is ready to fire.

Firing Pin Block

The firing pin block rests in the firing pin notch and prevents movement of the firing pin until the trigger is pulled. As the trigger is pulled, the firing pin block moves up and out of the firing pin notch. This movement allows a round to be fired when the hammer strikes the firing pin.

Half-Cock Notch

The half-cock notch stops the forward movement of the hammer during a mechanical failure.

Cycle of Operation

There are eight steps in the cycle of operation for the M9 service pistol.

Firing

Once the safety is off and the trigger is pulled to the rear, the hammer falls on the firing pin, which strikes the primer and ignites the round. See figure 1-5.

Figure 1-5. Firing.

Unlocking

As the slide assembly moves to the rear, the locking block rotates out of the notches in the slide. See figure 1-6.

Figure 1-6. Unlocking.

Extracting

As the slide moves rearward, the extractor withdraws the cartridge case out of the chamber. See figure 1-7.

Figure 1-7. Extracting.

Ejecting

As the face of the slide passes over the ejector, the case strikes the ejector and it is knocked upward and outward through the ejection port. See figure 1-8.

Figure 1-8. Ejecting.

Cocking

As the slide moves rearward, the hammer is pushed back, allowing the sear to engage the hammer hooks, cock the hammer to the rear, and place the pistol in the single-action mode. See figure 1-9.

Figure 1-9. Cocking.

Feeding

The slide starts forward, pushed by the recoil spring. The face of the slide makes contact with the cartridge at the top of the magazine, stripping it from the magazine and pushing it toward the chamber. See figure 1-10.

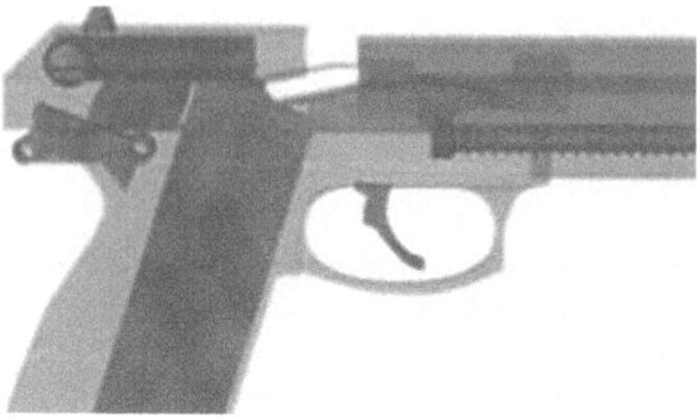

Figure 1-10. Feeding.

Chambering

As the slide continues forward, it pushes the cartridge into the chamber. See figure 1-11.

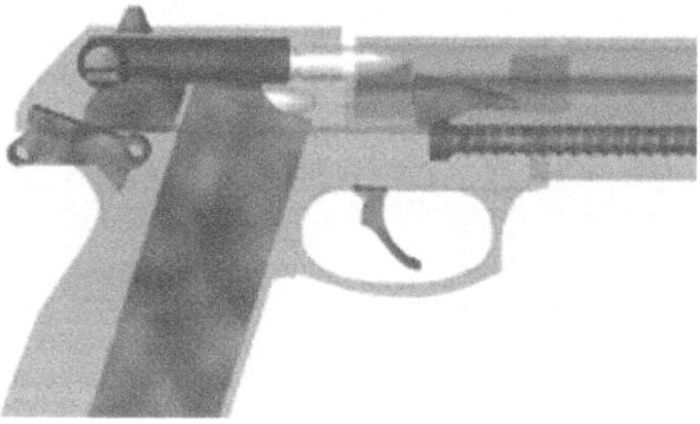

Figure 1-11. Chambering.

Locking

As the slide assembly continues to move forward, the locking block lugs move into the locking block recesses on the right and left sides of the slide. See figure 1-12 on page 1-6.

Figure 1-12. Locking.

Ammunition

The only ammunition authorized for the M9 service pistol is the NATO M882 9-mm ball. Dummy ammunition can be used during training (a dummy round has a hole drilled in its side and contains no primer). See figure 1-13.

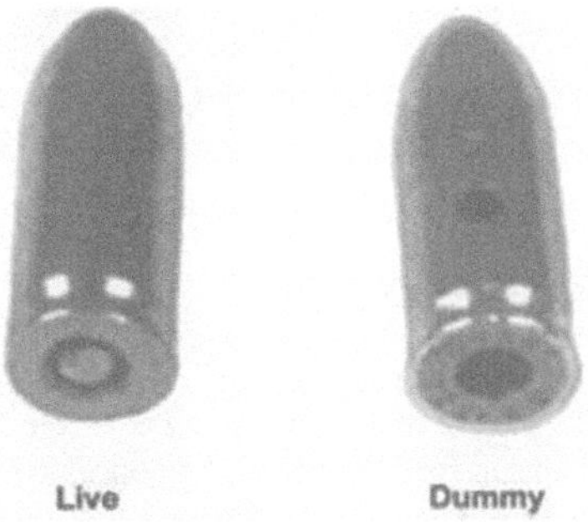

Figure 1-13. M9 Service Pistol Ammunition.

Do not open ammunition containers until the ammunition is to be used. Ammunition must be maintained in a high state of readiness. To care for ammunition—

- Keep ammunition dry and clean. If ammunition gets wet or dirty, wipe it off with a clean dry cloth.
- Wipe off light corrosion as soon as it is discovered. Never use ammunition that is heavily corroded, dented, or has the projectile pushed in.
- Do not expose ammunition to direct sunlight for long periods of time.
- Do not oil or grease ammunition. Dust or other abrasives can collect on greasy ammunition and may cause damage to the operating parts of the pistol. Oiled cartridges also produce excessive chamber pressure.

Wearing of the M9 Service Pistol's Gear

The proper placement of pistol gear helps ensure safety and aids the Marine in effectively handling and employing the pistol.

M12 Holster

The M12 holster consists of the holster, removable holster flap, and metal retaining clip. To check for proper placement of the holster, allow the right arm to hang freely. The holster should be slightly in front of the arm to permit easy access to the pistol upon presentation from the holster. See figure 1-14.

Figure 1-14. The M12 Holster and Ammunition Pocket.

Note: In most cases, the holster is issued with the holster flap installed for a right-handed Marine. To convert the holster for a left-handed Marine, remove the metal retaining clip and install the clip on the opposite side of the holster.

M1 Ammunition Pocket

The ammunition pocket attaches to the cartridge belt on the side opposite the holster in a position that best permits ready access for reloads. The magazine is stored in the ammunition pocket with the rounds down and pointed inboard.

M7 Shoulder Holster

The M7 shoulder holster consists of a holster with a thumb snap closure, shoulder strap, chest strap, and a belt retaining loop. The holster is positioned on the left side of the chest to provide easy and quick access with the right hand. See figure 1-15. The M7 holster comes fully assembled and has adjustable straps to accommodate each Marine's body size. The holster is available for right-handed Marines only, therefore, a left-handed Marine has to withdraw the pistol from the holster with the right hand and then transfer the pistol to the left hand before firing. (See chap. 8 for transferring the pistol from one hand to the other.) When the shoulder holster is worn properly—

Figure 1-15. M7 Shoulder Holster.

- The shoulder strap lays flat across the left shoulder with the shoulder pad directly on top of the shoulder.
- The chest strap attaches to the D-ring at the top of the holster and runs diagonally across the chest, underneath the right arm, and around the back where it attaches to the end of the shoulder strap.
- The belt retaining loop is at the bottom of the holster and attaches to the belt to stabilize the holster's position.
- The ammunition pocket attaches to the chest strap directly underneath the right arm.

Assault Holster

Some Marines (i.e., Marine security force, direct action platoon, and military police) are required to carry the assault holster. This holster has a retention strap that fastens over the top of the holster to retain the pistol. This type of holster generally has a thumb break on the retention strap that is disengaged to access the pistol. See figure 1-16.

Figure 1-16. Assault Holster.

Concealed Pistol Holster

Some Marines are required to carry a concealed pistol as part of their official duties. The primary consideration for placement of a concealed pistol holster is to ensure the pistol cannot be seen; therefore, the Marine must consider the type of clothing to be worn. A secondary consideration is to place the holster so the pistol can be presented easily. Typically, the best position for a concealed holster is just behind the strong side hip. This position best conceals the pistol while allowing it to be presented quickly. Another placement choice is in a shoulder holster, placing the pistol just under the weak side arm. See figures 1-17 through 1-20 on pages 1-8 and 1-9.

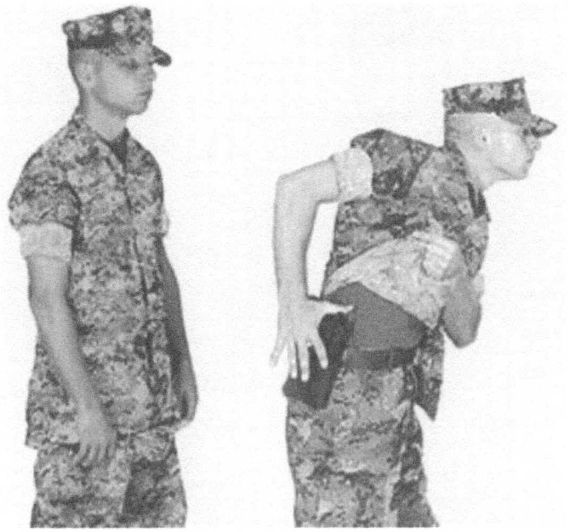

Figure 1-17. Wearing of the Concealed Pistol Holster—Utilities.

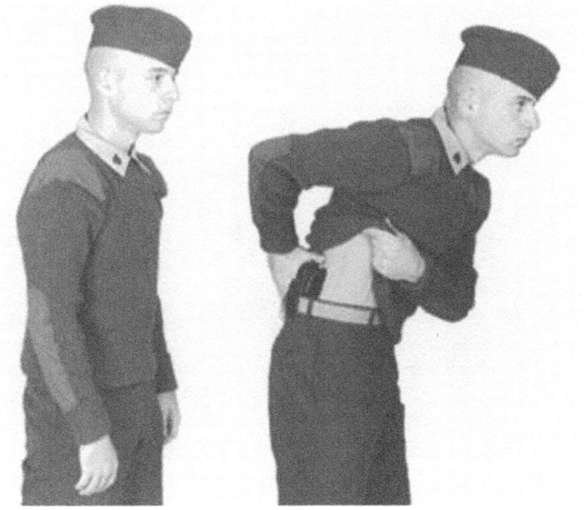

Figure 1-18. Wearing of the Concealed Pistol—Sweater.

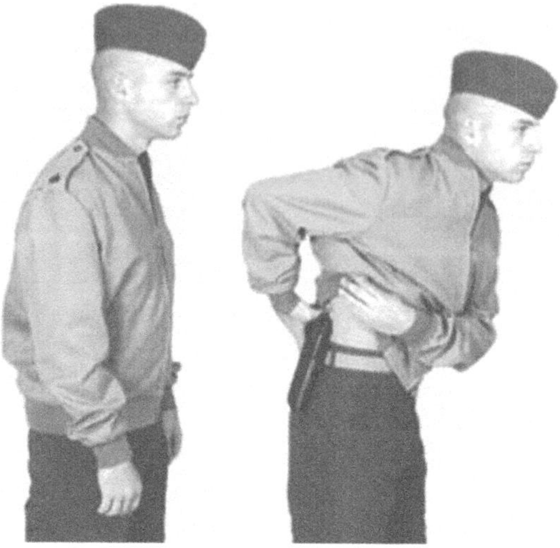

Figure 1-19. Wearing of the Concealed Pistol Holster—Jacket.

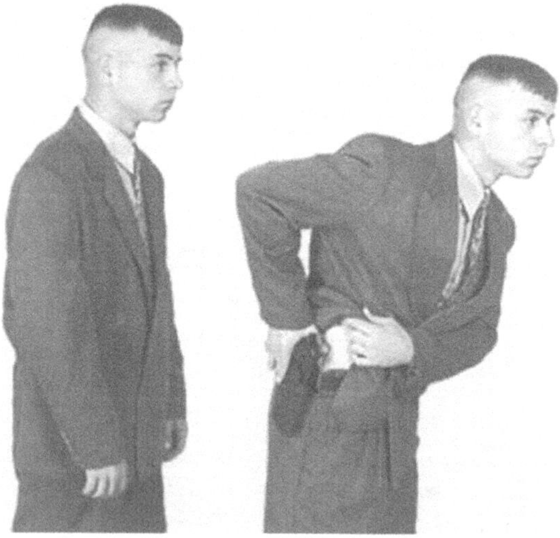

Figure 1-20. Wearing of the Concealed Pistol Holster—Civilian Attire.

Lanyard

The lanyard aids in pistol retention. It is issued in three sizes and is adjustable. The lanyard consists of a fabric cord, two cylindrical slip rings, and a metal clip that attaches the lanyard to the pistol's lanyard loop. See figure 1-21.

Figure 1-21. M9 Service Pistol (With Lanyard).

To don the lanyard—

- Adjust the slip rings so they are positioned flush with the lanyard's base (clip end).
- Place the right arm through the loop and place the loop over the head, resting on the left shoulder.
- Attach metal clip to the pistol's lanyard loop.
- Place the pistol in the holster.
- Use the left hand to hold the bottom slip ring against the base of the lanyard. Use the right hand to slide the top slip ring upward to position the loop of the lanyard under the arm. The lanyard should fit snugly against the body, but not restrict the Marine's movements.
- Tuck any excess cord behind the holster.
- Ensure that the lanyard is adjusted properly by removing the pistol from the holster and fully extending the right arm. The lanyard should be taut. Adjust as necessary.

  Note: Re-adjust the lanyard if any equipment changes are made (e.g., flak jacket).

Firing the M9 Service Pistol While Wearing Gloves

Not all combat engagements occur during ideal weather conditions. During cold weather, the Marine may find it necessary to wear gloves to protect the fingers from frostbite and help prevent stiffening of the hands. Gloves may also be worn in mission-oriented protective posture conditions. Gloves provide protection to the hands, however, they also may interfere with the Marine's ability to engage targets effectively. The added bulk of the gloves may affect the Marine's ability to manipulate the safety, magazine release button, magazine, hammer, and slide stop/release. For example, the Marine may need to exert more pressure to engage the magazine release or slide release buttons to compensate for the thickness of the gloves.

The principles of target engagement (see chap. 7) do not change while wearing gloves, however, the specific ability to manipulate and control the trigger is greatly affected by the thickness of the gloves around the fingers. Wearing gloves reduce the Marine's ability to feel, which makes it difficult to apply trigger control when firing. The Marine may find that more pressure than normal must be applied with the trigger finger just to establish initial contact with the trigger. Once the Marine can "feel" the trigger through the gloves, then the pressure required to fire a shot can be applied. This action may increase the chances of firing the pistol prematurely due to excessive pressure on the trigger. Dry firing while wearing gloves allows the Marine to learn how to apply trigger control consistently and determines how much pressure is needed to effectively fire a shot.

While wearing gloves, the Marine may find it difficult to fire the pistol in the double-action mode due to the position of the trigger and the limited amount of space between the trigger and the trigger guard. Therefore, if the situation permits, the Marine may wish to thumbcock the pistol to fire in single-action mode. In single-action mode, there is more space between the trigger guard and the trigger, making it easier to position the finger on the trigger. However, the Marine's ability to thumbcock the pistol may also be hindered by the gloves' bulk. Therefore, to thumbcock the pistol while wearing gloves, the Marine may perform one of the following methods:

- For method one, loop a section of 550 cord (approximately 2 inches) through the loop located on the top rear portion of the hammer. The length of the cord should not interfere with the pistol's cycle of operation or with the Marine's ability to establish sight alignment. Once the cord is attached to the hammer, take the pistol off safe and pull downward on the cord to cock the hammer. See figure 1-22.

Figure 1-22. Cocking the Pistol with 550 Cord.

- For method two, take the pistol off safe, rotate the pistol inboard, and place the top of the hammer against a secure surface (e.g., cartridge belt, table top, heel of boot). Apply pressure on the pistol to keep the hammer in place and push downward on the pistol in one continuous motion to cock the hammer. See figure 1-23.

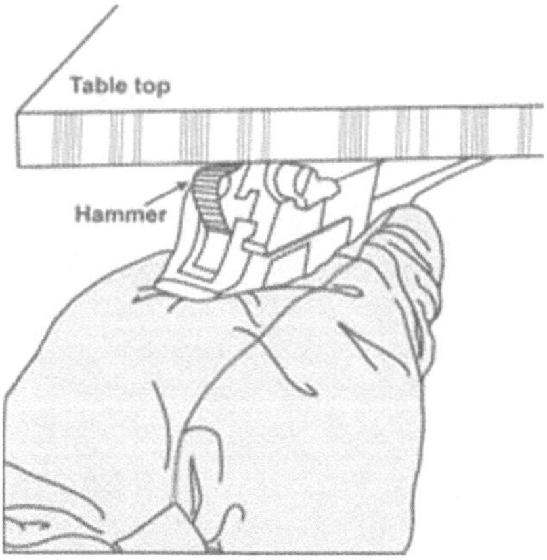

Figure 1-23. Cocking the Pistol on a Secure Surface.

Preventive Maintenance

If the M9 service pistol is to be effective, it must be maintained in a state of operational readiness at all times; therefore, maintenance of the M9 service pistol is a continuous effort. A clean, properly lubricated, well-maintained pistol will fire when needed.

Pistol Disassembly

Before disassembling the M9 service pistol, ensure that the pistol is in Condition 4. The pistol is in Condition 4 when the magazine is removed, the chamber is empty, the slide is forward, and the safety is on. To disassemble the pistol, perform the following steps in sequence:

- Hold the pistol in the right hand with the muzzle slightly elevated. Reach over the slide with the left hand and place the left index finger on the disassembly button and the left thumb on the disassembly lever. Press the disassembly button and hold it in place while rotating the disassembly lever downward until it stops.

Note: A left-handed Marine places the right thumb on the disassembly button and the right index finger on the disassembly lever.

- Pull the slide and barrel assembly forward and remove it while wrapping the fingers around the slide to hold the recoil spring and recoil spring guide in place.
- Turn the slide assembly over in the left hand until the recoil spring and recoil spring guide face up. Place the right thumb on the end of the recoil spring guide next to the locking block and compress the recoil spring and spring guide while lifting and removing them from the slide and barrel assembly. See figure 1-24. Allow the recoil spring to decompress slowly.

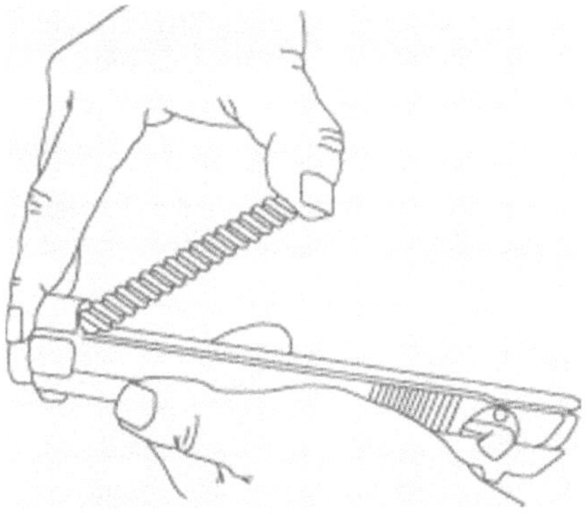

Figure 1-24. Removing the Recoil Spring and Recoil Spring Guide.

- Separate the recoil spring from the recoil spring guide.
- Push in on the locking block plunger with the right index finger while pushing the barrel forward slightly. Lift and remove the locking block and barrel assembly from the slide.

A Marine is not authorized to disassemble the pistol any further than the preceding steps. Any further disassembly must be performed by ordnance personnel. See figure 1-25.

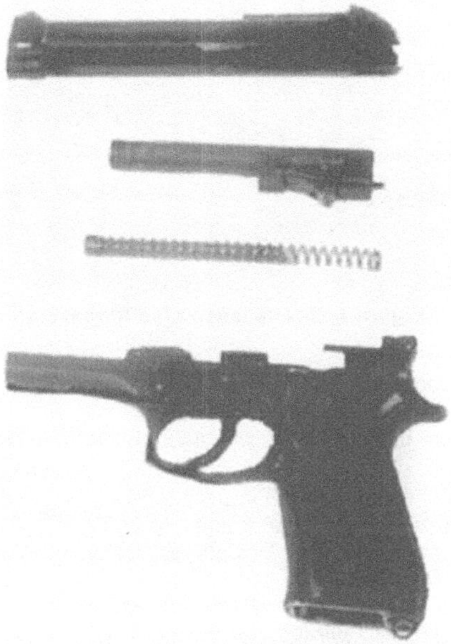

Figure 1-25. Disassembled M9 Service Pistol.

Disassembly of the Magazine

To disassembly the magazine, perform the following steps:

- Grip the magazine firmly in the left hand with the floorplate up and the thumb resting against the flat end of the floorplate.
- Release the floorplate by pushing down (with a blunt object; e.g., an ink pen) on the floorplate retainer stud in the center of the floorplate. At the same time, slide the floorplate a short distance forward with the thumb.
- Maintain the magazine spring pressure with the thumb and remove the floorplate from the magazine.
- Remove the floorplate retainer and magazine spring and follower from the magazine tube.

See figure 1-26.

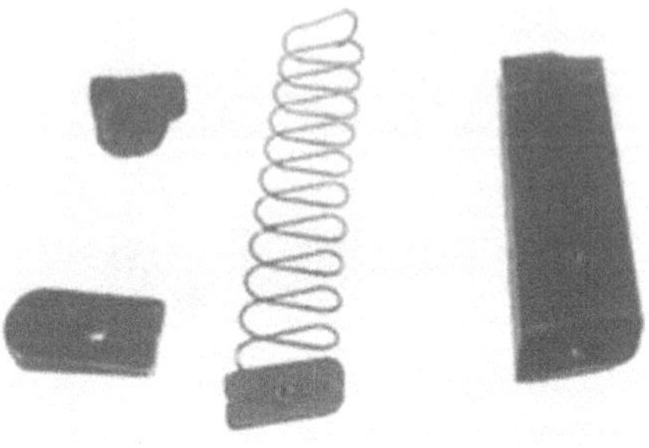

Figure 1-26. Disassembled Magazine.

Inspection of the Pistol

Once the pistol has been disassembled, it must be thoroughly inspected to ensure it is in a serviceable condition. Pistol inspection is continuous during the pistol's cleaning and reassembly:

Slide Assembly
- Check for free movement of the safety. Ensure the rear sight is secure.

Barrel Assembly
- Inspect the bore and chamber for pitting or obstructions.
- Check the locking block plunger for free movement of the locking block.
- Inspect the locking lugs for cracks and burrs.

Recoil Spring and Recoil Spring Guide
- Check the recoil spring for damage.
- Check that it is not bent.
- Check the recoil spring guide for straightness and smoothness.
- Check to be sure it is free of cracks and burrs.

Receiver Assembly
- Check for bends, chips, and cracks.
- Check for free movement of the slide stop and magazine catch assembly.
- Check the guide rails for excessive wear, burrs, cracks, or chips.

Magazine Assembly
- Check the spring and follower for damage.
- Ensure the lips of the magazine are not excessively bent and are free of cracks and burrs. The magazine tube should not be bent or dirty.

Cleaning and Lubricating the Pistol

Only authorized cleaning materials should be used to clean and lubricate the pistol. If these items are not issued with the pistol, they may be obtained from the armory. The following procedures are used to clean and lubricate the pistol—

Slide Assembly
- Clean the slide assembly with a cloth. A general purpose brush and cleaning lubricant protectant (CLP) can also assist in the removal of excess dirt and carbon buildup.
- Ensure the safety, bolt face, slide guides, and extractor are free of dirt and residue.
- Wipe dry with a cloth and apply a light coat of CLP.

Barrel Assembly
- Insert a bore brush with CLP into the chamber end of the barrel, ensuring that it completely clears the muzzle before it is pulled back through the bore.

Caution

Insert the bore brush through the chamber to prevent damage to the crown of the barrel.

- Repeat several times to loosen carbon deposits.
- Dry the barrel by pushing a swab through the bore.
- Repeat until a clean swab can be observed.
- Clean the locking block with a general purpose brush.
- Use the barrel brush to apply a light coat of CLP to the bore and chamber area and lubricate the exterior surfaces of the barrel and locking block.

Recoil Spring and Recoil Spring Guide
- Clean the recoil spring and recoil spring guide using CLP and a general purpose brush or cloth.
- Apply a light coat of CLP after wiping the recoil spring and recoil spring guide clean.

Receiver
- Wipe the receiver assembly clean with a cloth.
- Use a general purpose brush for areas that are hard to reach, paying special attention to the

disassembly lever, trigger, slide stop, hammer, and magazine release button.

- Apply a light coat of CLP.

Caution

Do not allow the hammer to fall with full force by pulling the trigger when the slide is removed. This can damage the receiver and hammer.

Magazine

- Clean the magazine tube and follower with CLP and a general purpose brush.
- Wipe the magazine spring, floorplate retainer, and floorplate clean with a cloth.
- Apply a light coat of CLP.

Pistol Re-assembly

After the M9 service pistol has been cleaned and lubricated, it must be properly re-assembled to ensure its serviceability. To re-assemble the M9 service pistol—

- Use the left hand to grasp the slide with the bottom facing up and the muzzle pointing toward the body. Use the right hand to grasp the barrel assembly with the locking block facing up. Use the index finger to push in the locking block plunger while placing the thumb on the base of the locking block.
- Insert the muzzle of the barrel assembly into the forward open end of the slide. At the same time, lower the rear of the barrel assembly by slightly moving the barrel downward. The locking block should fall into the notches of the slide assembly.
- Slip the recoil spring guide into the recoil spring.
- Insert the end of the recoil spring and recoil spring guide into the slide recoil spring housing. At the same time, compress the recoil spring and lower the spring guide until it is fully seated onto the locking block cutaway.
- Use the left hand to grasp the slide and barrel assembly, sights up, and wrap the fingers around the slide assembly to hold the recoil spring and guide in place. Align the slide assembly guide rails onto the receiver assembly guide rails.
- Push the slide rearward while pushing up on the slide stop with the thumb. Lock the slide to the rear while maintaining upward pressure on the slide stop. Rotate the disassembly lever upward. Listen for a click, an audible click indicates a positive lock.

Pistol Magazine Re-assembly

To re-assemble the magazine—

- Grip the magazine firmly in the left hand with the floorplate end up and the counting holes facing the Marine. Insert the follower into the magazine so the flat end of the follower is against the flat end of the magazine.
- Ensure the floorplate retainer is attached to the first curve of the bottom coil.
- Hold the spring upright with the right hand and insert the spring into the magazine tube so that the flat end of the floorplate retainer is against the flat end of the magazine.
- Push the magazine spring and floorplate retainer down with the right hand and hold it in place with the thumb of the left hand. Use the right hand to slide the floorplate over the side walls of the magazine until fully seated, which is indicated by an audible click.

Safety/Function Check

A safety/function check is performed after reassembling the M9 service pistol. Perform the following steps to ensure the pistol is operational:

- Ensure there is no ammunition in the chamber of the pistol.
- Ensure that the safety is in the safe position, then depress the slide stop, allowing the slide to return fully forward. At the same time, the hammer should fall to the full forward position.

- Pull and release the trigger. The firing pin block should move up and down but the hammer should not move.
- Place the safety in the fire position.
- Pull the trigger to check the double action. The hammer should cock and fall.
- Pull the trigger again and hold it to the rear. Use the fingers and thumb of the left hand to grasp the serrated sides of the slide just forward of the safety. Pull the slide to its rearmost position and release it while holding the trigger to the rear. Release the trigger, a click should be heard and the hammer should not fall.
- Pull the trigger to check the single action. The hammer should fall. Place the safety in the safe position.

If the safety/function check does not indicate an operational pistol, the Marine takes the pistol to organizational maintenance or the next authorized repair level.

User Serviceability Inspection

The Marine is responsible for performing a user serviceability inspection on the pistol prior to live fire. The user serviceability inspection ensures the pistol is in an acceptable operating condition. This inspection is not intended to replace the detailed pistol components inspection following disassembly or the limited technical inspection or pre-fire inspection conducted by a qualified armorer. To conduct a user serviceability inspection on the pistol, perform the following steps:

- Ensure the magazine release button is on the left side of the pistol for a right-handed Marine, the right side of the pistol for a left-handed Marine.
- Ensure the magazine seats into the magazine well when it is inserted and that it cannot be pulled out.
- Ensure the slide stays locked to the rear when the slide is pulled rearward with an empty magazine in the pistol.
- Ensure the magazine falls out freely when the magazine release button is depressed.
- Repeat the preceding four steps with the second magazine.
- Ensure the slide is locked to the rear before lubricating the spring guide, the top of the barrel just forward of the front sight, and the guide rails of the slide assembly behind the safety. With the muzzle pointed downward, work the slide several times and release.
- Inspect the pistol's external parts visually to ensure that there are no cracks or excessive wear.
- Perform a safety/function check of the pistol.

Pistol Maintenance in Adverse Conditions

Combat situations can place Marines in a variety of adverse conditions. Therefore, the M9 service pistol must be maintained properly to ensure its continued operation.

Extreme Cold

In extreme cold conditions, ensure that the following maintenance is performed:

- Clean and lubricate the pistol inside at room temperature, if possible.
- Apply a light coat of lubricant, arctic weather (LAW) to all functional parts.
- Always keep the pistol dry.
- Keep the pistol covered when moving from a warm to a cold area. This permit gradual cooling of the pistol and prevents freezing.
- Ensure that a hot pistol is not placed in snow or on ice.
- Keep snow out of the bore of the barrel. If snow should enter the bore, use a swab and cleaning rod to clean the bore before firing.

Hot, Wet Climates

Maintenance in hot, wet climates must be performed more frequently:

- Inspect hidden surfaces for corrosion. If corrosion is found, clean and lubricate.
- Remove handprints with a clean cloth in order to prevent corrosion.
- Dry the pistol with a cloth and lubricate it with CLP.
- Check ammunition and magazines frequently for corrosion. Disassemble and clean the magazines with CLP and wipe dry with a clean cloth. If necessary, clean ammunition with a dry cloth.

Hot, Dry Climates

In a hot, dry climate, dust and sand can cause stoppages and excessive wear on component contact surfaces during firing; therefore, keep the pistol covered whenever possible.

Corrosion is less likely to form on metal parts in a dry climate. Therefore, lightly lubricate internal working surfaces with CLP. Do not lubricate external parts of the pistol. Wipe excess lubricant from exposed surfaces. Do not lubricate internal components of the magazine.

Heavy Rain and Fording Operations

The following maintenance procedures are followed during periods of heavy rains or during fording operations:

- Always attempt to keep the pistol dry.
- Drain any water from the barrel prior to firing.
- Dry the bore with a swab and cleaning rod.
- Generously lubricate internal and external surfaces of the pistol with CLP.

Amphibious Conditions

If the pistol comes into contact with salt water, clean the pistol as soon as possible. If time does not permit cleaning in accordance with Technical Manual (TM) 1005A-10/1, *Operator's Manual, Pistol Semiautomatic, 9mm, M9*, then wash the pistol with fresh water

CHAPTER 2
WEAPONS HANDLING

Weapons handling is a method of providing consistent and standardized procedures for handling, operating, and employing the M9 service pistol. Understanding and applying the principles of weapons handling are critical to developing safe and consistent weapons skills. Strict adherence to training and diligent practice will make weapons handling instinctive. Mission accomplishment and survival during combat depend on a Marine's ability to react instinctively and with confidence.

Note: The procedures in this chapter are written for right-handed Marines; left-handed Marines should reverse directions as needed.

Safety Rules

The following safety rules are the foundation for responsible weapons handling. They must be observed at all times, both in training and combat.

Safety Rules

Rule 1: Treat every weapon as if it were loaded.
This rule is intended to prevent unintentional injury to personnel or damage to property from handling or transferring possession of a weapon.

Rule 2: Never point a weapon at anything you do not intend to shoot.
This rule enforces the importance of muzzle awareness and reinforces positive identification of the target.

Rule 3: Keep your finger straight and off the trigger until you are ready to fire.
This rule is intended to minimize the risk of firing the weapon negligently (when not firing, the trigger finger is straight along the receiver, outside of the trigger guard). This rule also reinforces positive identification of the target.

Rule 4: Keep weapon on safe until you intend to fire.
This rule enforces the use of the weapon's own safety feature and reinforces positive identification of the target.

Weapons Conditions

The M9 service pistol's level of readiness is defined by three specific conditions. The steps in the loading and unloading process take the pistol through the specific conditions that indicate the pistol's readiness for live fire. The Marine must understand and know the condition of his pistol at all times.

Condition 1. Magazine inserted, round in chamber, slide forward, and safety on.

Condition 2. Not applicable to the M9 service pistol.

Condition 3. Magazine inserted, chamber empty, slide forward, and safety on.

Condition 4. Magazine removed, chamber empty, slide forward, and safety on.

Determining a Weapon's Condition

There are two methods for determining the pistol's condition: checking the round indicator on the right side of the pistol and conducting a chamber check.

Checking the Round Indicator

When there is a round in the chamber, the upper surface of the extractor protrudes from the right side of the slide. The protrusion can be felt by sliding either the thumb or the index finger of the left hand over the top of the slide and across the extractor. See figure 2-1.

Conducting a Chamber Check

To conduct a chamber check—

- Point the pistol in a safe direction and grasp the pistol grip with the right hand.

Figure 2-1. Checking the Round Indicator.

- Place the trigger guard in the palm of the left hand. Use the thumb and index finger to grasp the forward end of the slide at the indentations under the front sight, behind the muzzle. See figure 2-2.

Figure 2-2. Placement of the Left Hand.

Caution

Ensure the muzzle does not cover the hand or fingers.

- Keep thumb in place around the pistol's backstrap and rotate fingers of the right hand over the top of the slide in front of the rear sight.
- Pull the slide to the rear by pushing forward with the right thumb and pulling back on the rear sight with the fingers. Use the left hand to steady the pistol and to assist in pulling the slide to the rear.
- Use the right hand to hold the slide to the rear (just enough to visually inspect the chamber for a round). Physically check for a round by inserting a finger of the right hand into the chamber area. See figure 2-3.

Figure 2-3. Chamber Check.

Note: At night or in low light conditions, the Marine's visibility is reduced; therefore, the Marine will have to rely on the physical check with the finger to determine if a round is in the chamber.

Caution

Pulling the slide too far to the rear while inspecting the chamber may cause a double feed or the ejection of a round.

- Remove the finger from the chamber and release tension on both hands to allow the slide to go forward. Ensure that the slide is all the way forward.

Weapons Commands

Weapons commands direct the Marine to safely load, unload, and employ the M9 service pistol. Six commands are used in weapons handling:

"*Load*"
This command is used to take a weapon from Condition 4 to Condition 3.

"*Make Ready*"
This command is used to take a weapon from Condition 3 to Condition 1.

"*Fire*"
This command is used to engage targets.

"*Cease Fire*"
This command is used to cease target engagement.

"*Unload*"
This command is used to take a weapon from any condition to Condition 4.

"*Unload, Show Clear*"
This command is used to require a second individual to check the weapon to verify that no ammunition is present before the weapon is put into Condition 4.

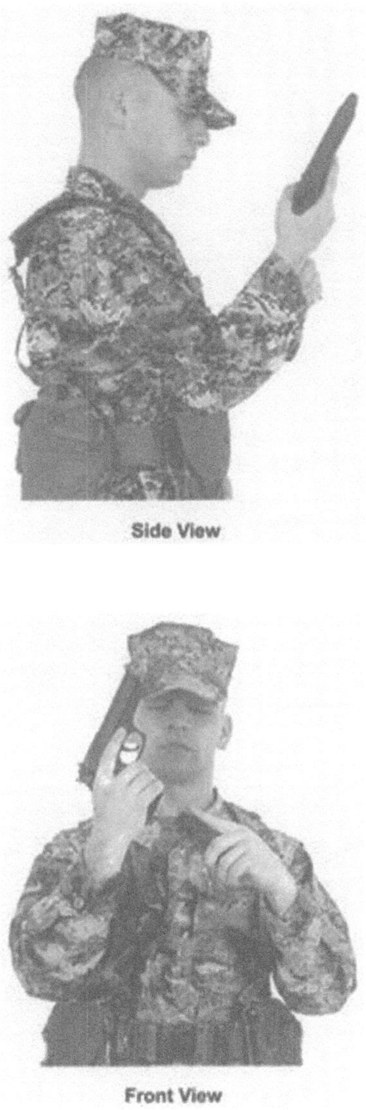

Figure 2-4. Indexing the Magazine.

Loading the Pistol

Perform the following steps to load the pistol (take the pistol from Condition 4 to Condition 3):

- Ensure the pistol is on safe.
- Use the right hand to grip the pistol grip firmly. Ensure that the pistol is pointed in a safe direction, bring the trigger guard to the right of eye level and cant the pistol so the magazine well faces inboard at approximately a 45-degree angle to the deck. Draw the right elbow in to facilitate control of the pistol.
- Use the left hand to remove a filled magazine from the ammunition pocket. Index the magazine by sliding the index finger along the forward edge of the magazine. See figure 2-4.
- Insert the filled magazine into the magazine well by guiding it with the index finger and, with the fingers extended, pushing it in with the heel of the hand until it is fully seated. Do not relinquish contact with the magazine until it is fully seated. See figure 2-5.

Figure 2-5. Seating the Magazine.

Making the Pistol Ready

Perform the following steps to take the pistol from Condition 3 to Condition 1:

- Firmly grip the pistol grip with the right hand. Ensure that the pistol is pointed in a safe direction and the slide is in its forward position.
- Rotate the magazine well outboard to facilitate pulling the slide to the rear. With the fingers and thumb of the left hand, grasp the serrated sides of the slide just forward of the safety. See figure 2-6.

Caution

Ensure the muzzle does not cover the hand or fingers.

- Pull the slide to its rearmost position by pushing forward with the right hand while pulling back on the slide with the left hand.
- Release the slide, this strips a round from the magazine and chambers it as the slide moves forward.
- Ensure the pistol remains on safe.
- Conduct a chamber check to ensure a round is in the chamber.

Note: A chamber check may be conducted at any time to check the pistol's condition.

Fire

Perform the following steps to fire the pistol:

- Keep trigger finger straight and use the right thumb to take the pistol off safe.
- Place the trigger finger on the trigger and apply pressure to the trigger until the shot is fired.

Right Side View

Left Side View

Figure 2-6. Grasping the Slide to Make Ready.

Cease Fire

Perform the following steps to execute a cease fire of the pistol:

- Remove the finger from the trigger and place it straight along the receiver.
- Place the pistol on safe without breaking the grip of the right hand.
- Assume a carry or transport position.

Unloading the Pistol

Perform the following steps to take the pistol from any condition to Condition 4:

- Use the right hand to grip the pistol firmly. Ensure that the pistol is on safe.
- Rotate the pistol so the magazine well is pointed inboard and angled down.

  Note: The angle of the magazine well must allow the magazine to fall freely from the well once the magazine release button is engaged.

- Depress the magazine release button to remove the magazine from the pistol. Catch the magazine with the left hand and retain it.
- Push upward on the slide stop with the right thumb and maintain pressure. Rotate the weapon so the chamber is outboard.

  Note: A left-handed Marine pushes upward on the slide stop with the left index finger.

- Reach over the top of the pistol with the left hand and grasp the slide serrations with the thumb and index finger. The left hand should partially cover the ejection port so it is positioned to catch an ejected round.
- Point the pistol in a safe direction and fully retract the slide and lock it to the rear. At the same time, catch the ejected round with the left hand. See figure 2-7.

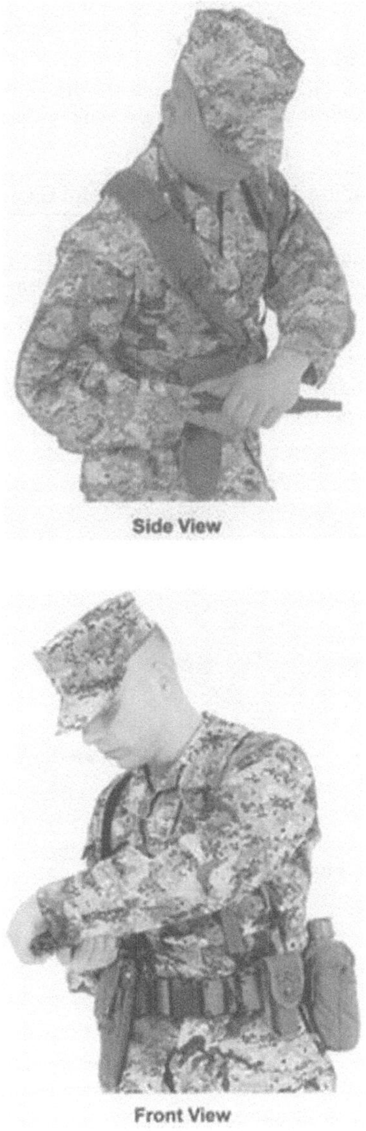

Figure 2-7. Catching the Ejected Round.

- Rotate the pistol so the inside of the chamber can be seen. Visually inspect the chamber to ensure it is empty.
- Press the slide stop to release the slide and observe it going forward on an empty chamber.

Unloading and Showing the Pistol Clear

Perform the following steps to take the pistol from any condition to Condition 4. See figure 2-8.

Figure 2-8. Unload, Show Clear.

- Use the right hand to grip the pistol firmly. Ensure that the pistol is on safe.
- Rotate the pistol so the magazine well is pointed inboard and angled down.

  Note: The angle of the magazine well must allow the magazine to fall freely from the well once the magazine release button is engaged.

- Depress the magazine release button to remove the magazine from the pistol. Catch the magazine with the left hand and retain it.
- Push upward on the slide stop with the right thumb and maintain pressure. Rotate the weapon so the chamber is outboard.

  Note: A left-handed Marine pushes upward on the slide stop with the left index finger.

- Reach over the top of the pistol with the left hand and grasp the slide serrations with the thumb and index finger. The left hand should partially cover the ejection port so it is positioned to catch an ejected round.
- Point the pistol in a safe direction and fully retract slide and lock it to the rear. At the same time, catch the ejected round with the left hand.
- Rotate the pistol so the inside of the chamber can be seen. Visually inspect the chamber to ensure it is empty.
- Bring the pistol to the administrative transport and have another Marine visually inspect the chamber to ensure that—
  - The chamber is empty, no ammunition is present, and the magazine is removed.
  - The pistol is on safe.
- Acknowledge that the pistol is clear.
- Press the slide stop to release the slide and observe it going forward on an empty chamber.

Emptying the Magazine

Once the pistol is unloaded, the pistol magazine can be emptied of ammunition. To empty the magazine, perform the following steps:

- Hold the magazine upright with the back of the magazine tube against the palm of the hand.
- Push the top round forward with the thumb and catch it with the other hand as it is removed.
- Repeat until the magazine is empty.

Filling the Magazine

Prior to loading the pistol, the pistol magazine must be filled with the prescribed number of rounds of ammunition. See figure 2-9. To fill the magazine, perform the following steps:

Figure 2-9. Filling the Magazine.

- Hold the magazine with the back of the magazine against the palm of the hand and the follower up.
- Use the other hand to place a round (primer end first) on the follower in front of the magazine lips.
- Press down on the round and slide the round completely back under the lips. The thumb or finger may push down on the back of the round to assist movement. The base of the round should be flush with the back of the magazine.

Repeat this procedure until the magazine is filled with the appropriate number of rounds. Holes on the back of the magazine allow the visual counting of rounds in five-round increments.

Reloading the Pistol

The Marine's ability to reload the pistol quickly improves his chance for success on the battlefield.

Dry Reload

A dry reload is conducted when the pistol runs out of ammunition during engagement and the slide locks to the rear. See figure 2-10. Perform the following steps to conduct a dry reload with the slide locked to the rear:

> *Note:* The pistol is not placed on safe during a dry reload; the trigger finger is taken out of the trigger guard and placed straight along the side of the receiver.

- Seek cover, if the situation permits.

Figure 2-10. Pistol Ran Dry of Ammunition.

- Retain the firing grip with the right hand and pull the pistol in close to the body to facilitate control. Bring the trigger guard to the right of eye level and cant the pistol so the magazine well is facing inboard at approximately a 45-degree angle to the deck. See figure 2-11.

Figure 2-11. Bring the Pistol Toward the Body.

Figure 2-12. Release Magazine and Unfasten Ammunition Pocket.

- Press the magazine release button and let the empty magazine fall to the deck. At the same time, unfasten the ammunition pocket to withdraw a filled magazine. See figure 2-12.

  Note: The primary objective during a dry reload is to get the pistol back in action as quickly as possible. Following engagement, retrieve the magazine before moving.

- Grasp the magazine by curling the middle finger and thumb of the left hand around the base of the magazine, with the index finger straight along the ammunition pocket. See figure 2-13.

Figure 2-13. Grasping the Magazine.

- Index the magazine: as the magazine is being withdrawn from the pocket, the index finger should be along the front of the magazine. See figure 2-14.

Figure 2-14. Indexing the Magazine.

- Rotate the hand up so the magazine is aligned with the magazine well.
- Glance quickly at the magazine well, insert the magazine into the magazine well (see fig. 2-15).

Figure 2-15. Glance at the Magazine Well.

- Bring the eyes back on target at the same time as the heel of the left hand seats the magazine, do not relinquish contact with the magazine. See figure 2-16.

Figure 2-16. Seating the Magazine.

- Roll both hands inward to establish a two-handed grip and press the slide release with the left thumb to allow the slide to move forward, chambering the first round, and present the pistol to the target. See figure 2-17.

Note: A left-handed Marine presses the slide release with his trigger finger.

Figure 2-17. Present Pistol Back to Target.

Figure 2-18. Facilitating Control of Pistol.

Condition 1 Reload

In a Condition 1 reload, a partially-filled magazine is removed from the pistol and replaced with a fully filled magazine. A Condition 1 reload is performed when there is a lull in the action or whenever deemed necessary by the Marine. To perform a Condition 1 reload—

- Retain the firing grip with the right hand and pull the pistol in close to the body to facilitate control. Keep the pistol pointed in the direction of the likely threat. See figure 2-18.
- Withdraw a filled magazine from the ammunition pocket with the left hand. Index the magazine and bring it up to the left of eye level. See figure 2-19.

Figure 2-19. Withdrawing and Indexing a Filled Magazine.

- Slide the index finger to the side of the magazine to grasp the magazine between the index and middle fingers. See figure 2-20.

Figure 2-20. Grasping a Filled Magazine.

- Raise the pistol and bring the trigger guard to the right of eye level and cant the pistol so that the magazine well is facing inboard at approximately a 45-degree angle to the deck.

  Note: The angle of the magazine well must allow the magazine to fall freely from the well once the magazine release button is engaged.

- Press the magazine release button with the right thumb to eject the partially-filled magazine from the magazine well. Grasp the magazine between the index finger and thumb. See figure 2-21.

Figure 2-21. Removing a Partially-Filled Magazine.

- Insert the filled magazine into the magazine well (see fig. 2-22). Use the heel of the hand to ensure it is fully seated (see fig. 2-23).

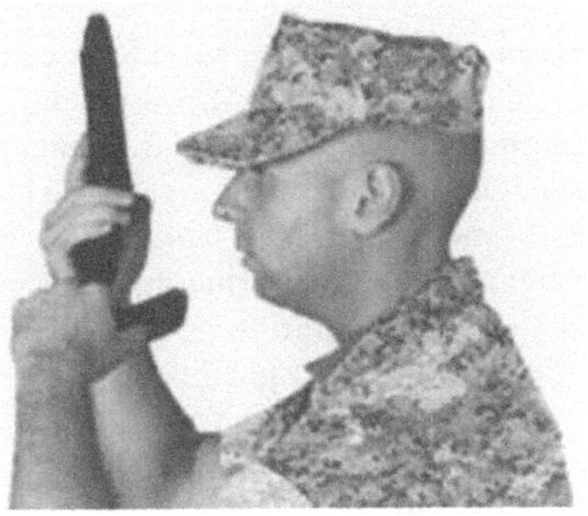

Figure 2-22. Inserting a Filled Magazine.

Figure 2-23. Seating a Filled Magazine.

- Lower the pistol and point it in the direction of the likely threat.

If time permits, examine the partially-filled magazine to determine the number of rounds remaining. Stow the partially-filled magazine in the ammunition pocket for later use.

Reloading Considerations

If possible, take cover before reloading. Always reload before leaving cover to take advantage of the protection.

When reloading, the first priority is to reload the pistol quickly so that it is ready to fire. During a reload, the Marine focuses on reloading only—not on the enemy.

The next priority is for the Marine to retain the magazine during the reload. However, the combat situation may dictate dropping the magazine to the deck when performing a reload (i.e., dry reload). If time permits (i.e., Condition 1 reload), the Marine picks the magazine up or secures the magazine (e.g., ammunition pocket, flak jacket) before moving to another location.

Remedial Action

The M9 service pistol is an effective and extremely reliable weapon. Proper care and preventive maintenance usually ensures the pistol's serviceability. However, stoppages, while infrequent, do occur. To keep the pistol in action, stoppages must be cleared as quickly as possible through remedial action. A malfunction cannot be corrected through remedial action by the Marine.

Stoppage

A stoppage is an unintentional interruption in the cycle of operation; e.g., the slide not moving forward completely. A stoppage is normally discovered when the pistol will not fire. Most stoppages can be prevented by proper care, cleaning, and lubrication of the pistol.

Many stoppages of the M9 service pistol are caused by shooter error. The Marine must be aware of shooter-induced stoppages in order to avoid them or to quickly identify and correct the stoppage and return the pistol to action. In a shooter-induced stoppage, the Marine—

- Fails to make ready.
- Fails to take the pistol off safe prior to firing.
- Engages safety while firing.
- Engages magazine release button while firing.
- Engages slide stop while firing (particularly prevalent when firing with an Isosceles grip).
- Fails to reset the trigger.
- Fails to recognize the pistol has run dry and the slide has locked to the rear.

Malfunction

A malfunction is a failure of the pistol to fire satisfactorily or to perform as designed (e.g., a broken front sight that does not affect the functioning of the pistol). A malfunction does not necessarily cause an interruption in the cycle of operation. When a malfunction occurs, the pistol must be repaired by an armorer.

Remedial Action

There is no one set of procedures (i.e., immediate action) that can be performed to clear all or even most of the stoppages that can occur with the M9 service pistol. Therefore, remedial action requires investigating the cause of the stoppage, clearing the stoppage, and returning the pistol to operation. When performing remedial action, the Marine should seek cover if the tactical situation permits. Once a pistol ceases to fire, the Marine must visually or physically observe the pistol to identify the problem before it can be cleared:

> *Note:* The steps taken to clear the pistol are based on what is observed.

- Remove the finger from the trigger and place it straight along the receiver.
- Bring the pistol in close to the body and in a position to observe the chamber.
- Pull the slide to the rear while observing the chamber area to identify the stoppage. See figure 2-24 on page 2-14.

> *Note:* Ensure the pistol does not move to safe when pulling the slide to the rear.

Figure 2-24. Observing Chamber.

- Correct the stoppage:
  - If there is a round in the magazine but not in the chamber (see fig. 2-25), the slide is released and a round is observed being chambered.

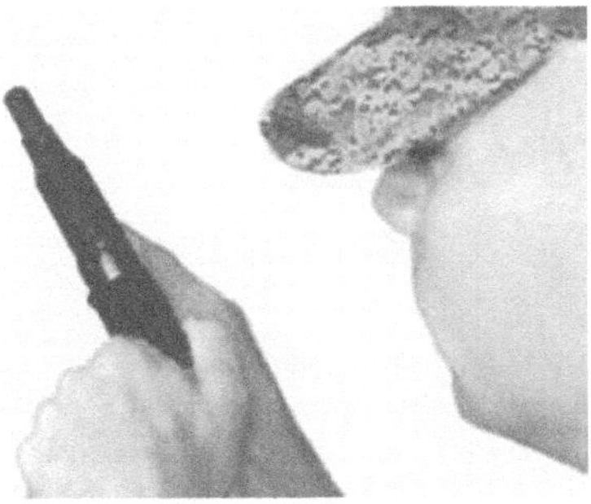

Figure 2-25. Round in Magazine but Not in Chamber.

  - If a round being chambered is not observed, the bottom of the magazine is tapped to seat it properly, and the slide is racked to the rear. See figure 2-26.
  - If there is no round in the magazine or chamber, a reload is conducted. See figure 2-27.

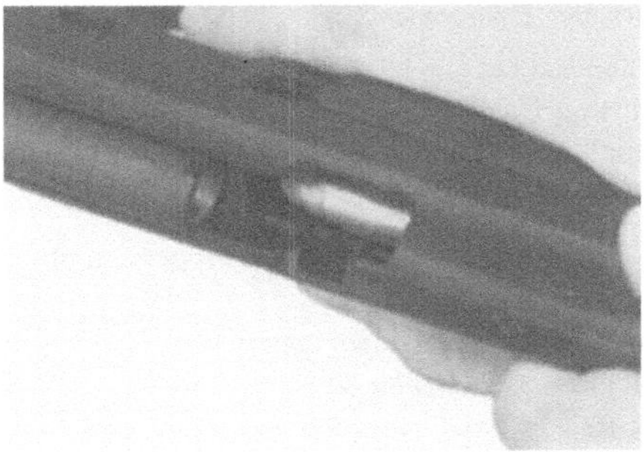

Figure 2-26. Round Not Being Chambered.

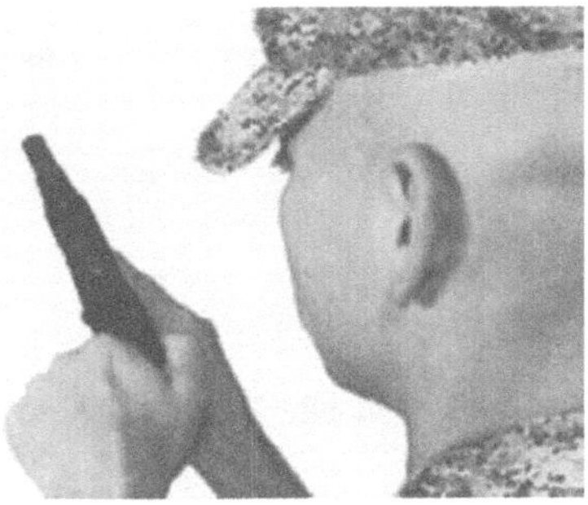

Figure 2-27. No Round in Magazine or Chamber.

- Fire the pistol.

Audible Pop or Reduced Recoil

WARNING

When an audible pop or reduced recoil is experienced, the Marine DOES NOT perform remedial action unless he is in a combat environment. An audible pop occurs when only a portion of the propellant is ignited. It is normally identifiable by reduced recoil and the pistol will not cycle. Sometimes, it is accompanied by excessive smoke escaping from the chamber area.

Training Environment

If an audible pop or reduced recoil is experienced during firing, cease fire immediately. Do not apply remedial action; instead, perform the following steps:

- Remove the finger from the trigger and place it straight along the receiver.
- Point the pistol down range.
- Place the pistol on safe.
- Raise a hand to receive assistance from available range personnel.

Combat Environment

The tactical situation may dictate correction of an audible pop or reduced recoil. To clear the pistol, perform the following steps:

- Remove the finger from the trigger and place it straight along the receiver.
- Seek cover if the tactical situation permits.
- Unload the pistol, but leave the slide locked to the rear.
- Insert something into the bore and clear the obstruction.
- Observe the barrel for cracks or bulges.
- Reload the pistol.

Weapons Carries

As the threat level increases, so should the Marine's readiness for engagement. Weapons carries are designed to place the Marine in a state of increased readiness as the threat level increases. There are two carries with the pistol: the Alert and the Ready. The carries permit quick engagement when necessary.

Alert

The Alert is used when enemy contact is likely (probable). See figure 2-28. The Marine performs the following steps to assume the Alert:

Figure 2-28. Alert.

- Ensure the pistol is on safe.
- Grasp the pistol grip firmly with two hands. The trigger finger is straight and the right thumb is on the safety and in a position to

- Extend the arms down at approximately a 45-degree angle to the body or bend the elbows. See figure 2-29.
- The muzzle of the pistol is pointed in the likely direction of the threat.

Ready

The Ready is used when there is no target, but contact with the enemy is imminent. The Marine performs the following steps to assume the Ready (see fig. 2-30):

- Ensure the pistol is on safe.
- Grasp the pistol firmly with two hands. The trigger finger is straight and the right thumb is on the safety and in a position to operate it.
- Extend arms and raise the pistol to just below eye level so a clear field of view is maintained.
- Point the muzzle of the pistol in the direction of enemy contact.

Weapons Transports

The M9 service pistol is transported in either the holster transport or the administrative transport.

Holster Transport

The holster transport is the most common method of carrying the pistol because it can be transported safely in the holster. This transport is used when there is no immediate threat (enemy contact is remote). See figure 2-31. To transport the pistol in the holster:

- Point the pistol in a safe direction.

WARNING

Ensure the pistol is pointed in a safe direction at all times and does not cover any part of the body while holstering.

Figure 2-29. Alert—Close Quarters.

Figure 2-30. Ready.

Figure 2-31. Holster Transport.

Figure 2-32. Administrative Transport.

- Ensure that the safety is on, the slide is forward, and the trigger finger is straight.
- Use the right hand to firmly grip the pistol grip and place the pistol in the holster:
  - Lift the flap of the holster with the left hand.
  - Look down at the holster, bring the pistol back to a position above the holster, and rotate the muzzle down into the holster.
- Push the pistol snugly into the holster and fasten the flap with the right hand.

Administrative Transport

The administrative transport is used to transport the pistol when the Marine does not have a holster. See figure 2-32. The Marine performs the following steps to assume the administrative transport:

- Establish a firm grip around the pistol grip with the right hand.
- Ensure the pistol is on safe, the magazine is removed, the slide is locked to the rear, and the trigger finger is straight along the receiver.
- Bend the elbow to approximately a 45-degree angle so the pistol is positioned near shoulder level. The wrist should be straight so the pistol's muzzle points up.

Combat Mindset

In a combat environment, the Marine must be constantly prepared to engage targets. When a target presents itself, there may be little time to react. The target must be engaged quickly and accurately. It is not enough to simply know marksmanship techniques, the Marine must be able to react instinctively.

The development of a combat mindset can be associated with the carries and holster transport for the pistol. The use of each carry/transport is dictated by the perceived level of threat. Each carry and transport should signify a stage of mental and physical preparedness for combat. The intensity of the Marine's mental and physical preparation depends on the likelihood of enemy contact.

Physical and Mental Preparation

Physical Preparation

In combat, targets can present themselves without warning. Therefore, it is essential for the Marine to maintain proper balance and control of the pistol at all times so that the pistol can be presented quickly and the target engaged accurately.

Speed alone does not equate to effective target engagement. The Marine should fire only as fast as he can fire accurately, never exceeding his physical ability to apply the fundamentals of marksmanship. To be effective in combat, the Marine must train to perfect the physical skills of shooting so they become second nature. The more physical skills that can be performed automatically, the more concentration that can be given to the mental side of target engagement.

Mental Preparation

While combat is unpredictable and constantly changing, the Marine can prepare mentally for the contingencies of the operational setting and confrontation with a threat. The stress of combat, coupled with the limited time available to engage targets, requires concentration on the mental aspects of target engagement; e.g., identification of targets, shoot/no-shoot decisionmaking, and the selection and use of cover. Minimizing stress and maximizing the limited time available to engage targets can be accomplished by—

- Knowing the combat environment and being constantly aware of the surroundings (e.g., terrain, available cover, possible threats) enables the Marine to quickly present the pistol and accurately engage targets.
- Identifying and evaluating possible courses of action and developing potential plans for target engagement that will be appropriate to the combat situation.
- Instilling confidence in a Marine's ability to fire well-aimed shots in the stress of a combat situation. A key factor in a Marine's level of confidence is the degree to which he has mastered the tactics, techniques, and procedures of pistol marksmanship.

Threat Levels

No Immediate Threat

When there is no immediate threat, the Marine assumes the holster transport. The pistol should be in Condition 1. This is the lowest level of awareness for the Marine in a combat environment, but the Marine must stay alert and aware of any nearby activity. To prepare for target engagement at this level, the Marine must—

- Be aware of likely areas of enemy contact.
- Be aware of the condition of his pistol.
- Establish a plan or course of action to present the pistol to a target should a target appear.
- Mentally review appropriate actions such as reloading and remedial action.

Contact Likely (Probable)

If enemy contact is likely (probable), the Marine assumes the Alert. When enemy contact is likely, the Marine should—

- Expect enemy contact and be constantly prepared to present the pistol.
- Search the entire area for indications of enemy targets and for suitable terrain features that offer cover and concealment. The Marine should avoid restricting the search to a single terrain feature because this hinders awareness to a sector of the battlefield and to approach by the enemy.
- Be mentally prepared for contact. Plan a course of action for immediate response to a target. Modify the plan of action as needed.

- Be physically prepared to fire. Maintain proper balance at all times. Avoid self-induced physical fatigue. For example, do not grip the pistol so tightly that fingers, hands, and arms tire from carrying the pistol.

Contact Imminent

When contact with an enemy target is imminent, assume the Ready. In this carry, the Marine is at the highest level of awareness and is constantly searching for and expecting a target. To fire well-aimed shots upon target detection, the Marine must be at the peak of his mental preparation, all distractions must be eliminated, and his focus must be on firing an accurate shot. In the Ready, the Marine must—

- Keep the pistol oriented in the general direction of observation (eyes, muzzle, target).
- Maintain a clear field of view above the pistol sights until the target is detected.
- Be mentally and physically prepared to engage the target. The Marine must be ready to:
  - Identify the target.
  - Sweep the safety.
  - Apply the fundamentals of marksmanship.
- Move only as fast as he is capable of delivering well-aimed shots, ensuring that speed of engagement does not exceed his physical abilities.
- Search the entire area for indications of enemy targets, lowering the pistol enough to observe a clear field of view of the area.

Transferring the Pistol

The Marine's ability to transfer a pistol to another Marine is critical to safe weapons handling. There are two methods for transferring the pistol from one Marine to another: show clear transfer and condition unknown transfer. Each transfer is performed based on the operational environment/ combat situation.

Show Clear Transfer

To transfer the pistol—

- Grasp the pistol firmly in the right hand while ensuring that the pistol is on safe.
- Remove and retain the magazine.
- Lock the slide to the rear and catch the round if there is a round in the chamber.
- Inspect the chamber visually to ensure it is empty and leave the slide locked to the rear. To transfer the pistol—
  - If the receiving Marine is to the right: Cradle the trigger guard in the palm of the left hand and wrap the fingers around the top of the pistol. Release the firing grip.
  - If the receiving Marine is to the left: With the left hand, grasp the slide of the pistol with the thumb over the slide and the fingers underneath. Release the firing grip.
- Ensure that the muzzle points up at a 45-degree angle in a safe direction and the chamber exposed. Hand the pistol to the other Marine, grip first. See figure 2-33.

Figure 2-33. Show Clear Transfer.

- The receiving Marine—
  - Grasps the pistol grip with the trigger finger straight along the receiver.
  - Inspects the chamber visually to ensure it is empty.
  - Ensures the pistol is on safe.

Condition Unknown Transfer

To transfer the pistol—

- Grasp the pistol firmly in the right hand while ensuring that the pistol is on safe. To transfer the pistol—
  - If the receiving Marine is to the right: Cradle the trigger guard in the palm of the left hand and wrap the fingers around the top of the pistol. Release the firing grip. See figure 2-34.

Figure 2-34. Condition Unknown Transfer to the Right.

  - If the receiving Marine is to the left: With the left hand, grasp the slide of the pistol with the thumb over the slide and the fingers underneath. Release the firing grip. See figure 2-35.

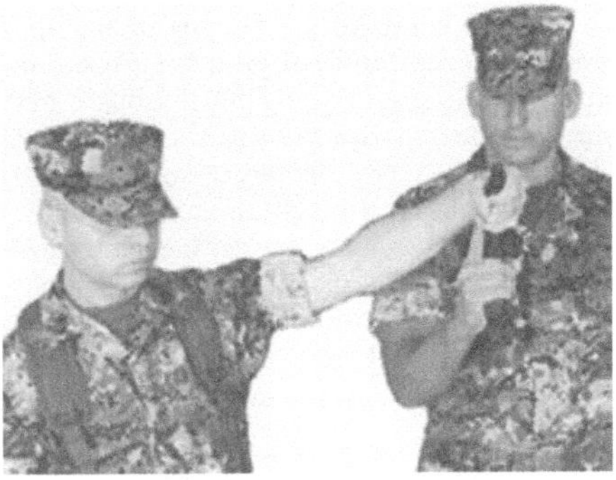

Figure 2-35. Condition Unknown Transfer to the Left.

- Ensure that the muzzle points up at a 45-degree angle in a safe direction. Hand the pistol to the receiving Marine, grip first.
- The receiving Marine—
  - Grasps the pistol grip with the trigger finger straight along the receiver.
  - Ensures the pistol is on safe.
  - Conducts a chamber check to determine the condition of the pistol.
  - Remove the magazine and count the number of rounds in the magazine by using the counting holes, if time permits. Re-insert the magazine into the magazine well ensuring it is fully seated.

Chapter 3
Fundamentals of Pistol Marksmanship

The fundamentals of pistol marksmanship are aiming, trigger control, and breath control. Understanding and applying the basic pistol marksmanship fundamentals ensures the Marine's effectiveness in target engagement. The fundamentals must be continually studied and practiced because they are the means by which accurate shots are placed on target. A Marine with a solid foundation in the fundamentals of marksmanship will be successful in the application of these fundamentals during combat.

Aiming

Maintaining the correct relationship between the pistol sights is essential for accurate target engagement. Because of the short distance between the pistol sights, a small error in their alignment causes a considerable error at the target.

Sight Alignment

Sight alignment is the relationship between the front sight and rear sight with respect to the aiming eye. Correct sight alignment is the front sight centered in the rear sight notch with the top edge of the front sight level **aligned** with the top edge of the rear sight. There should be equal space on either side of the front sight. See figure 3-1.

Figure 3-1. Sight Alignment.

Establishing Sight Alignment

The pistol is fired without benefit of bone support; therefore, the pistol is in constant motion. The Marine must understand this, yet continually strive to align the sights. To fire accurately, the sights must be aligned when the shot breaks.

Grip

The grip is key to acquiring sight alignment. If the grip is correct, the front and rear sights should align naturally. Dry fire during presentation of the M9 service pistol aids in obtaining a grip that allows sight alignment to be acquired consistently.

Controlled Muscular Tension

There must be enough controlled muscular tension in the grip, wrists, and forearms to hold the pistol steady and level the barrel to maintain sight alignment. Consistent tension stabilizes the sights and maintains sight alignment.

Sight Picture

Sight picture is the placement of the front sight in relation to the target while maintaining sight alignment. See figure 3-2 on page 3-2.

Because the pistol is constantly moving, sight picture is acquired within an aiming area that is located center mass on the target. The aiming area allows for movement of the sights on the target while maintaining sight alignment. Each Marine defines an acceptable aiming area within his own ability to stabilize the sights. Time, distance to the target, and personal ability affect dictate the aiming area. As the Marine becomes more proficient with the pistol, the aiming area becomes more precise.

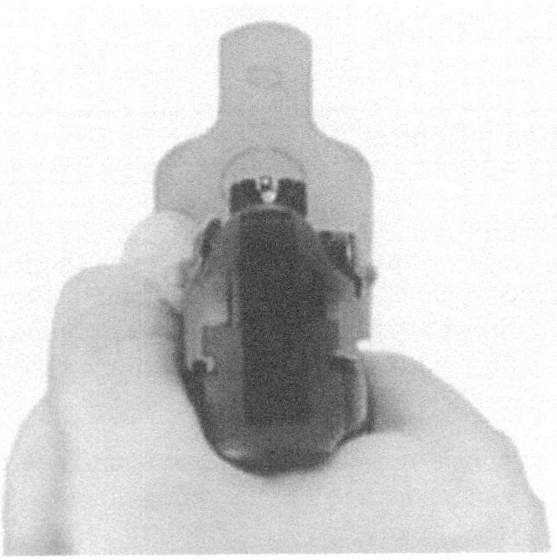

Figure 3-2. Sight Picture.

The aiming area is determined by the Marine's stability of hold. The proper grip stabilizes the sights so sight alignment can be maintained, but the sights move continuously within the aiming area of the target. The Marine understands the pistol's movement and learns to apply trigger control as he is obtaining sight alignment/sight picture within the aiming area so the shot breaks the moment sight picture is established.

Relationship Between the Eye and the Sights

The human eye can focus clearly on only one object at a time. The Marine must focus on the top edge of the front sight and fire the shot while maintaining the relationship between the front and rear sights within the aiming area. Focusing on the top edge of the front sight rather than the target keeps the front sight clear and distinct, which allows the Marine to detect minor variations in sight alignment. Secondary vision allows the Marine to see the target (although slightly blurred) and maintain sight picture within his aiming area.

Trigger Control

Trigger control is the Marine's skillful manipulation of the trigger that causes the pistol to fire while maintaining sight alignment and sight picture. Proper trigger control aids in maintaining sight alignment while the shot is fired.

Sight Alignment and Trigger Control

Aiming and trigger control are mutually supportive—one cannot be performed without the other. Sight alignment and trigger control must be performed simultaneously to fire an accurate shot. As pressure is applied to the trigger, the sights may move, causing them to be misaligned. To fire accurate shots, the sights must be aligned when the shot breaks. Trigger control can actually assist in aligning the sights. With proper trigger finger placement and consistent muscular tension applied to the grip, the sights can be controlled as the trigger is moved to the rear. If the sights move extensively while pressing the trigger, this can indicate an improper grip or inconsistency in the muscular tension being applied to the grip.

Grip

A firm grip is essential for good trigger control. The grip is established before applying trigger control and is maintained throughout the firing process. To establish the grip, the hand is placed around the pistol grip in a location that allows the trigger finger to move the trigger straight to the rear while maintaining sight alignment. Once the grip is established, it should be firm enough to allow manipulation of the trigger while maintaining sight alignment. The pressure applied to the grip must be equal to or more than the pressure required to move the trigger to the rear. If the pressure is not applied correctly, the sights move as the trigger is pressed to the rear and sight alignment is disturbed.

Trigger Finger Placement

Once the grip is established, the finger is placed on the trigger. Placement of the finger should be natural and allow free movement of the trigger finger. A natural trigger finger placement allows the trigger to be moved straight to the rear while maintaining sight alignment. If the finger presses the trigger to the side, it can cause an error in sight alignment and shot placement.

Each Marine must experiment with finger placement in order to determine effective placement on the trigger. Once established, effective trigger finger placement allows the trigger to be consistently moved straight to the rear while maintaining sight alignment.

Types of Trigger Control

Uninterrupted Trigger Control

During uninterrupted trigger control, the Marine applies a steady, unchanging pressure to the trigger until the shot is fired. Uninterrupted trigger control is particularly effective at close range, when the target area is large, and when stability of hold is not critical for accuracy. To apply uninterrupted trigger control, apply pressure on the trigger while maintaining focus on the top edge of the front sight. Continue pressure on the trigger to begin moving the trigger straight to the rear while obtaining sight alignment and sight picture. Move the trigger straight to the rear in a single, smooth motion with no hesitation.

Interrupted Trigger Control

Interrupted trigger control is particularly effective at longer ranges, when the target is small, and when stability of hold is critical to maintaining sight picture in the aiming area. This method is also used if the pistol sights move outside the aiming area when applying trigger control. If the sight picture is outside the aiming area, the Marine stops and holds the rearward movement on the trigger until sight picture is re-established. When sight picture is re-established, the rearward movement of the trigger is continued until the shot is fired.

Breath Control

Breathing causes movement of the chest, abdomen, and shoulders, which causes the pistol sights to move vertically while attempting to aim and fire. Therefore, it is necessary to stop breathing for a period of time while firing a shot or a series of shots.

The object of breath control is to stop breathing just long enough to fire the shot while maintaining sight alignment, stabilizing the sights, and establishing the sight picture. To be consistent, the breath should be held at the same point in the breathing cycle; i.e., the natural respiratory pause.

Breathing should not be stopped for too long because it has adverse visual and physical effects. Holding the breath longer than is comfortable results in a lack of oxygen that causes vision to deteriorate and then affects the ability to focus on the sights.

Application of Marksmanship Fundamentals in Field Firing

Compression of the Fundamentals

Pistol engagements typically occur over close distances and are short in duration. Because an immediate response to the threat is required, the application of the fundamentals must be a conditioned response that is executable in a compressed time. The goal of successful, quick target engagement is the application of the fundamentals of marksmanship while firing the shot the moment weapon presentation is complete.

The time required to move the trigger to the rear while acquiring and maintaining sight alignment and sight picture is unique to each Marine and is based on his capabilities. Each Marine should know his abilities and fire only as quickly as he is capable of firing accurately. The Marine must not exceed his shooting skills in an effort to get rounds off quickly.

Aiming

In field firing, the fundamentals are applied in a compressed time so sight alignment and sight picture are achieved as the shot is fired. Although the target must be quickly engaged in combat, sight alignment is still the first priority: strive for a clear front sight. Distance to the target and the size of the target affects sight alignment as follows:

- As the distance to the target increases and the size of the target decreases, sight alignment becomes more critical to target engagement. Accurate sight picture/sight alignment cannot be compromised for speed.
- Sight alignment is critical to the effective engagement of smaller targets such as partially exposed targets.
- As the distance to the target decreases, perfect sight alignment is not as critical, but there must be a relationship between the sights and their placement on the target within the aiming area to ensure accuracy.

Trigger Control

Proper trigger control aids in maintaining sight alignment while the shot is fired. As pressure is applied to the trigger, the sights may move, causing them to be misaligned. Therefore, the Marine is constantly re-aligning the sights as pressure is applied to the trigger. Sight alignment and trigger control must be performed simultaneously to fire an accurate shot.

Breath Control

During combat, the Marine's breathing and heart rate often increase due to physical exertion or the stress of battle. The key to breath control in field firing is to stop breathing just long enough to fire an accurate shot or a series of shots.

CHAPTER 4
PISTOL FIRING POSITIONS AND GRIP

The M9 service pistol is fired from the standing, kneeling, and prone positions. Each firing position may be adapted to either a Weaver or Isosceles variation, each possessing a distinct advantage in combat. The Weaver variation stabilizes the pistol sights. The Isosceles variation manages recoil. The advantages apply whether the Marine is firing in the standing, kneeling, or prone position. The Marine must select and assume a stable firing position that provides a solid foundation for accurate shooting while meeting the demands of the combat situation

Note: The procedures in this chapter are written for right-handed Marines; left-handed Marines should reverse directions as needed.

The Pistol Firing Position

During combat, the Marine selects a firing position based on mobility, observation of the enemy, and stability.

Mobility

A firing position must provide mobility should the Marine need to move. The standing position permits maximum mobility because it is quickly assumed and easily maneuvered from and it permits lateral mobility to engage widely dispersed targets. The prone position provides limited mobility because it is the most time-consuming position to get into and out of and it lacks the lateral mobility required to engage dispersed targets.

Observation of the Enemy

A firing position should allow observation of the enemy while minimizing the Marine's exposure. During combat there can be many obstructions to a clear field of view. Terrain features such as vegetation, earth contours, and manmade structures often dictate the firing position. The prone position normally allows the least exposure, but it usually provides a limited field of view. Kneeling may provide a wider field of view, but generally provides less concealment.

Stability

A solid firing position establishes a stable foundation for target engagement and provides accurate and consistent shooting. The definition of a stable position is one in which the body is positioned so as to resist forces that cause motion (i.e., recoil and movement of the pistol sights). The standing position is the least stable firing position, while the prone is the most stable firing position. A consistent, stable position is assumed for two distinct purposes:

- Minimize the pistol's movement in order to control the sights and to deliver accurate fire on a target.
- Minimize the affects of recoil in order to recover the sights to the same area on the target.

During combat, it may be necessary to engage the same target more than once to eliminate it. If the firing position is stable, the pistol's sights should recover to the same area on the target, allowing rapid re-enactment. Distributing the body's weight to balance the position also stabilizes the pistol and allows better management of recoil. A pistol firing position is stabilized through controlled muscular tension. Because the pistol is fired without benefit of bone support, muscular tension is needed in the body to stabilize the position and the pistol sights. Muscular tension must follow these guidelines—

- A consistent amount of muscular tension is needed to hold the pistol steady so the sights are aligned with the aiming eye and the target.

- Controlled and consistent tension in the body provides resistance that the Marine uses to manage recoil and bring the sights back on target quicker. However, too much tension can cause strain or trembling.
- Muscular tension is correct when the Marine can control the pistol before, during, and after firing the shot.

Pistol Firing Grip

A proper grip is one that provides maximum control of the pistol before, during, and after firing. It stabilizes the pistol sights before firing, allows trigger control to be applied during firing, and manages recoil after firing.

Before Firing: Stabilize the Sights

To fire an accurate shot, the sights must be stabilized prior to and as the bullet exits the muzzle of the pistol. A proper grip controls the alignment of the pistol's sights and stabilizes the sights so an accurate shot may be fired. The following guidelines apply to the establishment of a proper grip—

- There must be muscular tension in the wrist and forearms. Consistent muscular tension in the wrist, forearms, and grip helps maintain sight alignment by reducing the movement in the grip that can cause movement in the sights. The grip is correct when it allows the sights to be naturally aligned to the aiming eye.
- When establishing a two-handed grip, equal pressure must be applied with both hands. Consistent, equal pressure from both hands stabilizes the sights and allows them to be aligned and level with respect to the aiming eye.
- The hand grips as high on the back strap as possible, bringing the centerline of the bore as close as possible to a parallel line with the bones in the forearm.

During Firing: Allow Trigger Control

The Marine's grip provides a foundation for the movement of the trigger finger. The trigger finger applies positive pressure on the trigger as an independent action, completely free of the other muscles of the gripping hand. The Marine should not apply excessive pressure to the web of the hand touching the pistol's back strap because excessive pressure interferes with the manipulation of the trigger by the trigger finger.

After Firing: Manage Recoil

Once a shot is fired, the pistol recoils, disturbing alignment of the sights. A proper grip facilitates a quick recovery from recoil so the sights quickly return to the same area on the target. The Marine's grip determines the following during recoil:

- The amount the muzzle climbs during recoil depends on the amount of controlled muscular tension in the grip and wrists applied to stabilize the pistol and create consistency in resistance to recoil. Controlled muscular tension allows the sights to recover consistently back on target within a minimum amount of time.
- Firm, equal pressure must be applied to the grip with both hands to ensure that the pistol does not slip during recoil. An improper grip or lack of controlled muscular tension causes the pistol to move after the shot is fired, disrupting sight alignment and requiring the Marine to re-establish his grip.
- Locking out the wrist of the firing hand, similar to executing a punch, helps provide resistance to recoil and speed recovery. The elbows should be at a position slightly less than locked out to help absorb the recoil and aid in recoil management.

Withdrawing the Pistol From the Holster

The firing grip is not established in the holster, however, a proper firing grip can be assumed quickly if the pistol is withdrawn from the holster correctly. The pistol should be withdrawn from the holster in one continuous, fluid motion:

- Place the heel of the left hand at the center of the torso with the fingers extended toward the target. At the same time, unfasten and release the D-ring with the right hand. See figure 4-1.

Figure 4-1. Withdrawing the Pistol From the Holster—Step 1.

Note: The left hand is placed on the torso in a position that allows a two-handed firing grip to be established in a minimum amount of movement.

- Place the right thumb on the forward edge of the lower portion of the holster and the fingers around the back edge of the holster, keeping the trigger finger straight. See figure 4-2.

Note: The left hand may assist in holding the holster flap up.

Figure 4-2. Withdrawing the Pistol From the Holster—Step 2.

- Slide the hand up the holster until the fingers come in contact with the pistol grip. At the same time, keep the thumb above the pistol to guide the holster flap up. See figure 4-3 on page 4-4.
- Grasp the pistol grip with the fingers and draw the pistol straight up. Continue withdrawing the pistol while moving the thumb to a position on the safety. See figure 4-4 on page 4-4.

Note: This hand placement allows a firing grip to be established once the thumb disengages the safety. Any adjustments made to the firing grip after the safety is disengaged should be minor.

Figure 4-3. Withdrawing the Pistol From the Holster—Step 3.

Figure 4-4. Withdrawing the Pistol From the Holster—Step 4.

- Once the muzzle clears the holster, rotate the muzzle forward to clear the body. Ensure the muzzle is pointed in a safe direction.
- Establish a two-handed grip on the pistol by joining the left hand with the right hand in the front of the torso. See figure 4-5.

Caution

Ensure the muzzle does not cover the left hand when establishing the two-handed grip.

Figure 4-5. Withdrawing the Pistol From the Holster—Step 5.

Weaver and Isosceles Variations

The size of the target, distance to the target, time, and type of engagement needed (i.e., two shots, single precision shot) determine whether to fire in the Weaver or Isosceles variation. The Weaver and Isosceles variations each consist of three firing positions: standing, kneeling, and prone. The firing position chosen is based on the combat situation and the Marine's body configuration, and it must permit balance, control, and stability during firing. In addition, there is a specific firing grip that supports each position and the combination of the grip and the body configuration is key to establishing the variation. See table 4-1 for a list of advantages and disadvantages of both the Weaver and the Isosceles variations.

At longer ranges, the target is smaller, a more precise shot is required to eliminate the target, and any small movement moves the sights off the target; therefore, the pistol must be steadied. Although the Weaver variation is effective at any distance, some aspects of the position make it more effective for long range or precision shots on small or partially exposed targets; i.e., sight alignment and sight picture are easier for stability of hold.

The Isosceles variation is effective at any distance, however, some aspects of the position make it more effective for close range engagements. When confronted with a target, the natural physical reaction is to face the target and push out with the arms. This makes the Isosceles variation advantageous for quick engagements at close range. When a target is at close range, it must be engaged quickly before it engages the Marine. The management of recoil is a bigger factor in close-range engagements because it is more likely that multiple shots will be fired to eliminate the target and the sights have to recover quickly back on target. However, at close ranges, the target is larger so stability of hold is not as important because it is easier to hold the sights on target and sight picture is not as critical.

Standing Position

The standing position is the most often employed position during a pistol engagement due to the short distance of the engagement and the nature of combat. When properly assumed, the standing position provides a stable base for firing, a clear field of view, and excellent mobility. The standing position can be adapted to either the Weaver variation or the Isosceles variation.

Weaver Standing Position

The key to successful employment of the Weaver variation is the body's angle to the target and the push-pull pressure applied to the grip. See figure 4-6.

Table 4-1. Advantages and Disadvantages.

Variation	Advantages	Disadvantages
Weaver	Additional balance, control, and stability of hold during firing due to placement of the arms (left arm bent, pistol is in close to the body). Easier to maintain sight picture because the pistol's foundation is steady.	Recoil has a greater impact due to the hand placement on the pistol (some of the pistol grip is exposed and pressure is applied in two different directions around the pistol). Recovery of the sights back on target may take longer since recoil is affected.
Isosceles	Recoil has a lesser impact because muscular tension and grip pressure are evenly distributed around the pistol. Allows quicker recovery of the sights on target.	Stability of hold is degraded due to the pistol being further from the body without support. Since it is harder to steady the pistol, acquiring sight picture becomes more difficult.

To assume the Weaver standing position—

- Face the target and make a half turn to the right, keeping the pistol oriented toward the target. This orients the body at approximately a 40 to 60 degree angle oblique to the target. The shoulders are angled to the target, the left shoulder forward of the right. The feet are about shoulder-width apart, the left foot forward of the right.
- Grip the pistol grip firmly with the right hand. Place the right thumb on the safety.
- Keep the shoulders at a 40 to 60 degree angle oblique to the target and raise the right arm and extend it across the body toward the target. Ensure that the right shoulder does not roll forward or turn toward the target.
- Extend the left arm to the target, bending the left elbow to join the left and right hands. The left elbow is inverted and tucked in toward the body so the left arm supports the pistol.

Figure 4-6. Weaver Standing Position.

Note: The angle of the body determines how much the elbow bends.

- Establish a two-handed firing grip in the Weaver variation. See figure 4-7.
- Place the palm of the left hand over the front of the right hand so the palm covers the curled fingers of the right hand. The trigger guard should rest in the "V" formed by the left thumb and forefinger. The knuckles of the left hand are just outboard of the trigger guard. A portion of the pistol grip is exposed.
- Rest the trigger finger naturally, straight and outside of the trigger guard, so the finger can be moved quickly and easily to the trigger.
- Rest the left thumb against the receiver so that both thumbs are on the left side of the pistol. Once the safety is disengaged with the right thumb, the left thumb is placed over the right thumb and positive pressure is applied to hold the right thumb in place.
- Apply rearward pressure with the left hand and forward pressure with the right hand to achieve a "push-pull" grip. Isometric tension (push-pull) stabilizes the pistol during firing.

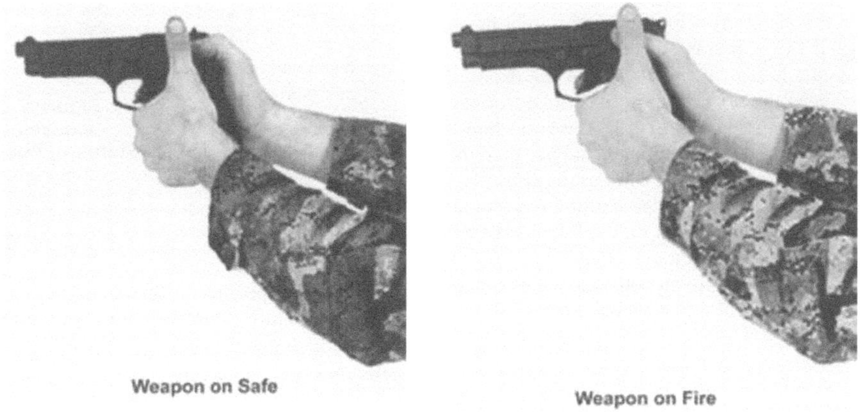

Figure 4-7. Weaver Grip.

- Lean forward slightly and apply muscular tension throughout the body to stabilize the position and manage recoil. The muscular tension in the upper body is not symmetrical due to the "push-pull" tension applied on the grip.
- Keep the head erect so the aiming eye can look through the sights.

Isosceles Standing Position

The key to the Isosceles variation is that the body is squared to the target and equal pressure is applied on the pistol from the grip. To assume the Isosceles standing position—

- Face the target with feet approximately shoulder width apart. The shoulders are squared to the target.
- Establish a two-handed firing grip in the Isosceles variation. See figure 4-8.

Figure 4-8. Isosceles Grip.

- Grip the pistol grip firmly with the right hand. Place the right thumb on the safety.
- Place the heel of the left hand on the exposed portion of the pistol grip in the pocket formed by the fingertips and heel of the right hand. There should be maximum contact between the pistol grip and the hands. Wrap the fingers of the left hand over the fingers of the right hand. Ensure both thumbs rest on the left side of the pistol and point toward the target.
- Apply equal pressure on both sides of the pistol to allow for the best management of recoil.

Note: Ensure the left thumb does not apply excessive pressure to the slide stop or slide.

- Rest the trigger finger naturally, straight and outside of the trigger guard, so the finger can be moved quickly and easily to the trigger.

Notes: Index finger of the left hand may or may not rest on the front of the trigger guard.

Marines with large hands must ensure that their right thumb does not rest on the slide stop, preventing reliable pistol operation.

- Elevate and extend the arms toward the target.
- Roll the shoulders forward and shift the body weight slightly forward to stabilize the position and better manage recoil. The left foot may be slightly forward of the right foot to balance the position. There should be an equal amount of muscular tension on both sides of the body to best manage recoil.
- Tuck the head between the shoulders; the head is extended forward but kept erect so the aiming eye can see through the sights. See figure 4-9.

Figure 4-9. Isosceles Standing Position.

Kneeling Position

The kneeling position offers a smaller exposure than the standing position and greater stability. Increased stability makes the kneeling position effective for longer range shooting. It does not, however, offer as much mobility for quick reaction as the standing position. The kneeling position can be quickly assumed and allows firing from various types of cover. Depending on the cover and the need for observation, the kneeling position may be adapted to a high kneeling, a medium kneeling, a low kneeling, or a two-knee kneeling position. The kneeling positions can be adapted to either the Weaver variation or the Isosceles variation.

Weaver Kneeling Position

The advantage of the Weaver variation of the kneeling position is that it provides bone support due to the left elbow's placement on the knee. The Weaver variation further enables firing from the side of cover while exposing less of the body to a threat. To assume the Weaver kneeling position, the following steps are basic to all adaptations:

- Make a half turn to the right, drop the right foot back or step forward with the left foot, and place the right knee on the deck. The body is positioned at a 40 to 60 degree angle oblique to the target.
- Blade the shoulders at a 40 to 60 degree oblique angle to the target, the left shoulder forward of the right.
- Extend the arms toward the target.
- Bend forward at the waist to better manage recoil.
- Place the flat part of the upper left arm, just above the elbow, in firm contact with the flat surface formed on top of the bent knee. The point of the left elbow extends just slightly past the left knee. However, depending on the need for stability or observation of the enemy, the elbow does not have to rest on the knee.

The following steps provide specifics for each adaptation:

- High kneeling: the toes of the right foot are curled and in contact with the deck or the inside of the foot may be in contact with the deck. Depending on the need for observation, the buttocks may or may not rest on the right heel. The left leg is bent at the knee; the shin straight up and down. The left foot is flat on the deck. See figure 4-10.

Figure 4-10. Weaver High Kneeling.

- Medium kneeling: the right ankle is straight with the foot stretched out and the bootlaces in contact with the deck. The left leg is bent at the knee, the left foot flat on the deck. The right shin may be angled to the body to create a tripod of support for the position. See figure 4-11.

Figure 4-11. Weaver Medium Kneeling.

- Low kneeling: the right ankle is turned so the outside of the foot is in contact with the deck and the buttocks are in contact with the inside of the foot. The right shin may be angled to the body to create a tripod of support for the position. See figure 4-12.

Figure 4-12. Weaver Low Kneeling.

- Two-knee: drop both knees onto the deck. The toes may be curled to get into and out of the position quickly. Depending on the need for observation of the enemy, the buttocks may or may not rest on the heels. See figure 4-13.

Isosceles Kneeling Position

The advantage of the Isosceles variation is that it enables the Marine to fire over the top of cover while exposing less of the body to a threat. To assume the Isosceles kneeling position, the following steps are basic to all adaptations:

- Drop the right foot back or step forward with the left foot and place the right knee on the deck.
- Square the shoulders to the target.
- Extend the arms toward the target.
- Lean forward with the shoulders rolled forward and the head tucked between the shoulders to better manage recoil.

Figure 4-13. Weaver Two-Knee Kneeling.

The following steps provide specifics for each adaptation:

- High kneeling: the toes of the right foot are curled and in contact with the deck or the inside of the foot may be in contact with the deck. Depending on the need for observation, the buttocks may or may not rest on the right heel. The left leg is bent at the knee, the shin is straight up and down. The left foot is flat on the deck. See figure 4-14.

Figure 4-14. Isosceles High Kneeling.

- Medium kneeling: the right ankle is straight with the foot stretched out and the bootlaces in contact with the deck. The left leg is bent at the knee, the left foot is flat on the deck. The right shin may be angled to the body to create a tripod of support for the position. See figure 4-15.

Figure 4-15. Isosceles Medium Kneeling.

- Low kneeling: the right ankle is turned so the outside of the foot is in contact with the deck and the buttocks are in contact with the inside of the foot. The right shin may be angled to the body to create a tripod of support for the position. See figure 4-16.

Figure 4-16. Isosceles Low Kneeling.

- Two-knee: drop both knees onto the deck. The toes may be curled to get into and out of the position quickly. Depending on the need for observation of the enemy, the buttocks may or may not rest on the heels. See figure 4-17.

Figure 4-17. Isosceles Two-Knee.

Prone Position

The prone position is easily assumed, stable, and presents a small target to the enemy. Since the prone position places most of the body on the deck, it offers great stability for long range shooting. However, it is the least mobile of the firing positions and may restrict the field of view. The prone position can be adapted to either the Weaver variation or the Isosceles variation.

Weaver Prone Position

The Weaver variation of the prone position produces a cocked leg position by angling the body to the target and cocking the leg to support the position. The Weaver prone position is ideal for firing from behind cover. See figure 4-18.

Figure 4-18. Weaver Prone Position.

To assume the Weaver prone position, perform the following steps:

- Face the target and make a half turn to the right (this places the body at a 40 to 60 degree oblique to the target). Grip the pistol in the right hand, placing the pistol in a position that facilitates control of the weapon. Ensure the pistol is pointed in a safe direction and does not cover any portion of the body.
- Move the body to the deck by using either the squat or drop method, keeping the body at a 40 to 60 degree oblique to the target.

  Squat Method
  - Squat down and place the left hand on the deck.
  - Kick both feet backward and come down on the right side of the body with the right arm extended toward the target.
  - Ensure the pistol does not cover the body or the left hand. See figure 4-19.

  Drop Method
  - Drop to a kneeling position.
  - Place the left hand on the deck in front of the body, push the pistol out toward the target.
  - Roll the right side of the body onto the deck.
  - Ensure the pistol does not cover the body or the left hand. See figure 4-20 on page 4-12.
- Bring the left knee up to support the firing position and to raise the diaphragm off the deck so as not to interfere with breathing. The inside of the knee rests on the deck. The knee is drawn up to provide maximum stability for the position.
- Establish a two handed-firing grip on the pistol.

Figure 4-19. Squat Method.

- Place the left elbow on the ground for stability. For maximum stability, strive to keep the grip firmly placed on the deck.
- The head may rest against the right arm so the pistol sights can be aligned. The head may be canted as long as the aiming eye can look directly through the sights. Strive to keep the pistol sights as level as possible while acquiring sight alignment.

To make minor increases in elevation, keep the left hand in place and firmly on the deck and raise the right hand to achieve the desired elevation. However, contact between the right and left hands must be maintained to stabilize the pistol. See figure 4-21 on page 4-12. (There is a tradeoff between obtaining the needed elevation and losing stability, so the Marine must strike a balance between the two.)

Figure 4-20. Drop Method.

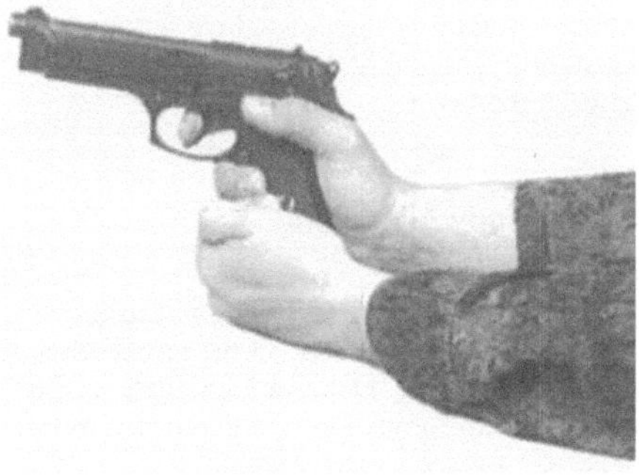

Figure 4-21. Increasing Elevation (Minor Adjustments).

Isosceles Prone Position

The Isosceles variation of the prone position produces a straight leg position. See figure 4-22.

To assume the Isosceles prone position:

- Stand facing the target. Grip the pistol in the right hand, placing the pistol in a position that facilitates control of the weapon. Ensure the pistol is pointed in a safe direction and does not cover any portion of the body.
- Move the body to the deck by using either the squat or drop method.

  Squat Method
  - Squat down and place the left hand on the deck.
  - Kick both feet backward and come down on the right side of the body with the right arm extended toward the target.
  - Ensure the pistol does not cover the body or the left hand. See figure 4-23.

  Drop Method
  - Drop to a kneeling position.
  - Place the left hand on the deck in front of the body.

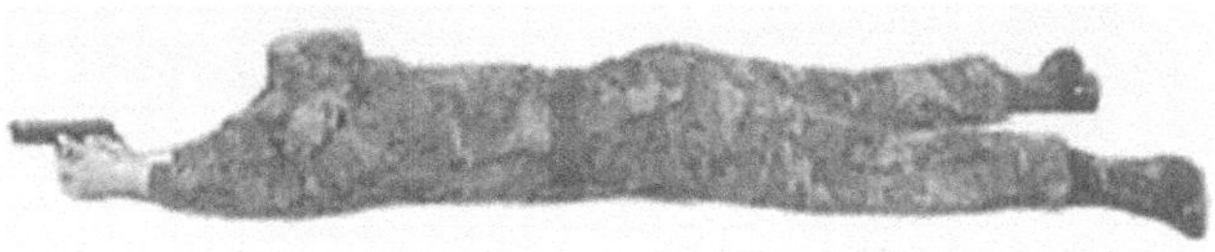

Figure 4-22. Isosceles Prone Position.

Figure 4-23. Squat Method.

- Push the pistol out toward the target, and roll the right side of the body onto the deck.
- Ensure the pistol does not cover the body or the left hand. See figures 4-24.

- Establish a two-handed firing grip on the pistol.
- Spread the legs to a position that provides maximum stability. The insteps of both feet may be flat on the deck or the toes may be curled and dug into the deck.
- Keep the pistol sights as level as possible while acquiring sight alignment. Keep the head in a position to allow the aiming eye to look directly through the sights.
- When wearing a helmet, the head may be canted slightly and rest against the right arm to push the helmet from the eyes so the sights can be aligned. Likewise, the pistol may be canted outboard to allow the aiming eye to look directly through the sights.

To make minor increases in elevation, keep the left hand in place and firmly on the deck and raise the right hand to achieve the desired elevation. However, contact between the right and left hands must be maintained to stabilize the pistol. See figure 4-25 on page 4-14. (There is a tradeoff between obtaining the needed elevation and losing stability, so the Marine must strike a balance between the two.)

Natural Body Alignment

The body must be properly aligned to the target so the sights fall naturally on the target when the pistol is presented. It takes a combination of body alignment and consistent muscular tension to

Figure 4-24. Drop Method.

Figure 4-25. Increasing Elevation (Minor Adjustments).

ensure the sights fall naturally to the same area of the target every time the pistol is presented. The Marine can execute the following to check natural body alignment and to ensure the sights center on the aiming area:

- Orient the body to a target and establish a variation of the standing position and a two-handed firing grip on the pistol. Aim in on the target.
- Close the eyes and take a deep breath.
- Open the eyes and see where the pistol sights are in relation to the target. If the pistol sights are right or left of the target, move the feet to adjust the position right or left.

  Note: Do not force the sights onto the target area by moving the arms; this increases the muscular tension on one side of the body, disturbs balance, and makes recoil harder to manage.

If the sights are significantly out of alignment when the pistol is at eye level, it may be an indication of a poor grip. When the grip is correct (to include the muscular tension in the grip, wrist, and forearms), the sights should align to the point that only minor adjustments are needed to align the sights to the aiming eye.

Repeat the preceding steps as necessary. Body alignment and muscular tension are correct when the sights are naturally placed in the same area on the target every time the Marine aims on the target.

CHAPTER 5
USE OF COVER AND CONCEALMENT

On the battlefield, a firing position that allows maximum observation of the enemy as well as cover and concealment is a necessity. A good position provides a solid foundation for the pistol, maximizes the use of cover to provide protection from enemy fire, allows mobility, and provides observation of the enemy. Where possible, the cover should be used to provide additional support for the position.

When contact with the enemy is made, it is important to seek cover as quickly as possible. Cover is anything that provides protection from enemy fire. Cover should be, at a minimum, thick enough to stop small arms fire and high enough to protect the Marine when firing from behind it. The effective use of cover enables engagement of enemy targets while affording protection from enemy fire. Cover can also be effectively used to conceal the Marine from enemy view while searching for targets.

Concealment is anything that hides a Marine from enemy view; however, it might not afford protection. Concealment can be provided by brush, trees, etc.

> *Note:* The procedures in this chapter are written for right-handed Marines; left-handed Marines must reverse directions as needed.

Cover Materials

Natural cover (rocks, logs, rubble, etc.) is best because it is hard to detect. But any material (including buildings, structures, etc.) that may protect an individual from small arms fire can be used for cover. Some common types of cover material are as follows.

Dirt

The best type of cover is dirt packed to a minimum thickness of 18 inches.

Cinder Blocks

Cinder blocks used as foundations for houses or walls can be used for cover, but they can be penetrated. They are also brittle and can shatter upon impact from small arms fire, which can cause injury by secondary fragmentation.

Trees, Logs, and Telephone Poles

Wood is a relatively dense material; therefore, it offers good covering protection since bullets have a tendency to fragment as they penetrate. Live trees have a greater resistance to bullet penetration than dead wood. Wood that has been treated with creosote, such as telephone poles and railroad ties, offers better protection from projectiles than untreated wood, but it still does not ensure protection from small arms fire.

Sandbags

Sandbags can be used for cover. However, they should always be packed tightly and tamped down to increase their density. If loosely filled or moist, a bullet can more easily penetrate the sandbag. Doubling or overlapping sandbags also increases their protective qualities.

Considerations for Firing From Cover

Adjusting the Firing Position

Cover can provide additional support for the firing position. The firing position is adjusted to ensure stability, mobility, and observation of the enemy.

The Marine chooses a firing position based on his height in relation to the height of the cover. The firing position must minimize exposure to the enemy but allow observation of the area.

Although the cover provides additional support, the Marine continues to apply the same amount of muscular tension in the grip, wrist, and forearms. Muscular tension is still necessary to stabilize the pistol sights and manage recovery.

Because the sights are higher than the muzzle of the pistol, the Marine must ensure that the muzzle of the pistol clears the cover as he obtains sight alignment/sight picture on the target. The closer the Marine is to the cover, the easier it is to ensure the muzzle clears the cover.

Applying the Weaver and Isosceles Variations

The type of cover may dictate which variation of a firing position will be the most effective. The position should provide the Marine with the maximum amount of stability and control and allow the Marine to manage recoil effectively in order to recover on target.

Weaver Variation

The Weaver variation exposes less of the body from behind cover due to the angle of the body. The Weaver variation may be better suited for firing from behind the right or left side of cover. For example, the Weaver prone is ideal for firing from behind a log.

Isosceles Variation

The Isosceles variation is good for firing over the top of cover (e.g., a window). The Isosceles prone is ideal for firing from behind narrow cover (e.g., a telephone pole).

Keeping the Body Behind Cover

Avoid inadvertent exposure of any part of the body. Be especially aware of the top of the head, elbows, knees, or any other body part that may extend beyond the cover.

Log or Curb

When firing from behind a log or curb, the Marine must present the lowest possible silhouette and may use the log or curb for maximum support of the position. For maximum protection, muzzle clearance is kept as close as possible to the top of the log or curb. The Marine may fire from either the side or the top of the log, depending on cover and concealment (see fig. 5-1). The Marine fires over the top of a curb when it is used for cover.

Figure 5-1. Firing Around a Log.

Wall or Barricade

Firing is done from either the side or over the top of a wall or barricade. See figure 5-2.

Figure 5-2. Firing From a Barricade.

Window

If the Marine has not been detected by the enemy, he should use the side of the window or the window sill for support. Ideally, it is best to fire from the corner of the window sill when using the window for support. If there is little chance for detection or the shot can be made without support, the Marine should remain back and to the side of the window opening so the pistol does not protrude and his body is concealed by the shadows/darkness of the room. If the Marine is positioned too close to the window, his body provides a silhouette to the enemy.

Vehicle

In many combat situations, particularly in urban environments, a vehicle may be the best form of cover. When using a vehicle for cover, the engine block provides the most protection from small arms fire. The Marine establishes a position behind the front wheel or front door jamb so the engine block is between him and the target. See figure 5-3.

Figure 5-3. Firing Over the Hood of a Vehicle.

From this position, the Marine fires over the hood of the car or underneath the car from behind the wheel. At the back of the car the axle and the wheel provide the only cover. If the Marine must shoot from the back of the car, he must position himself directly behind the wheel as much as possible. See figure 5-4.

Figure 5-4. Firing From the Back of a Vehicle.

Providing Support for the Position and the Pistol

Support helps stabilize both the firing position and the pistol and enable the Marine to maintain sight alignment and sight picture.

The forearms or hands can contact the support to stabilize the pistol. The Marine may rest the pistol on or against the support as long as the support does not interfere or affect the pistol's cycle of operation. See figure 5-5.

Figure 5-5. Hand Resting on Support.

When firing over the top of cover, the Marine can establish a supported position and stabilize the position by resting the trigger guard or the magazine on the cover. The pistol may be pushed up against the support so the "V" formed by the receiver and the front of the trigger guard rests firmly against the support. See figure 5-6.

Figure 5-6. Trigger Guard Resting Against Support.

When firing around the sides of cover, the Marine can establish support and stabilize the position by placing the back of the hand or arm against the cover. Avoid placing the slide of the pistol against the cover because it can interfere with the pistol's cycle of operation. However, the pistol can be canted and placed against the cover so the trigger guard or the "V" formed between the receiver and trigger guard rests against the support. This position enables the Marine to expose less of himself to the enemy. See figure 5-7.

Figure 5-7. Hand Resting Against Side of Support.

When using a vehicle for cover, the Marine can establish additional support for the pistol by positioning himself in the car behind the door jamb (frame of door) and placing his hands or pistol against the "V" formed by the open door and door frame. See figure 5-8.

When shooting from the left side of cover, the Marine still uses his right hand and eye. See figure 5-9. He may have to cant his head and the pistol to the left to establish sight alignment. For right-handed Marines, shooting from the left side of cover may expose more of the Marine to the enemy than shooting from the right side.

Figure 5-8. Firing From Behind the Door Jamb of a Vehicle.

Figure 5-9. Firing From the Left Side of Cover.

Changing Positions

If the Marine has been firing from cover and has to reload or clear a stoppage behind cover he should attempt to resume firing from a different position. The enemy is aware of the Marine's current position and will be ready to engage him once he re-appears.

Moving

In combat, the Marine must be constantly aware of his surroundings and the available cover should enemy contact occur during movement from one position to another.

When moving from cover to cover, the Marine selects the next cover location and plans the route of movement before leaving his present position. This is done by quickly looking from behind cover to ensure the area is clear, ensuring the head and eyes are exposed for as short a time as possible. If necessary, the Marine should conduct a Condition 1 reload before moving from cover. Once the Marine is committed to moving, he must focus on the move until cover is reassumed.

Supported Firing Positions

During combat, the Marine may not have the time to assume a perfect firing position. He must know instinctively that his position is correct rather than follow a regimented sequence of movements to ensure its correctness. With training, the Marine can assume stable firing positions quickly and instinctively by incorporating the use of cover for support. Support provides foundations for the firing position; which, in turn, provides support for the pistol. To maximize the support provided by the position, the firing position should be adjusted to fit or conform to the shape of the cover.

A supported firing position minimizes exposure to the enemy, maximizes the stability of the pistol and protection from fire, and provides observation of the enemy. Any stable support may be used (e.g., logs, sandbags, walls). The surrounding environment dictates the support and position.

The size, distance to the target, and time affect the need for stability and recovery in the selection of a supported firing position. For example, if the target is a great distance from the Marine, he may sacrifice some of his ability to manage recoil in order to assume a supported firing position that provides him the additional stability needed to fire accurately at long range. But, recovery may be more important for the Marine if he must fire multiple shots on target quickly; therefore, he may sacrifice some stability in his supported firing position in order to engage a target with multiple shots.

Supported Prone

The supported prone position presents the lowest silhouette and provides maximum protection from enemy fire. The supported prone position can be assumed behind a tree, a wall, or almost any type and size of cover. It is flexible and allows firing of the pistol from all sides. To assume the supported prone position and maximize the use of cover, the position is kept as low as possible to ensure no part of the body is exposed to the enemy. If the cover is narrow, the Marine positions his body directly behind cover and keeps his legs together. The Marine's body is in line with the pistol and directly behind the pistol. The Isosceles prone position presents a smaller target to the enemy and more body mass to absorb recoil. The Weaver prone position is ideally suited to fire from around cover (e.g., log) because of the angle of the body, but this position presents a larger target to the enemy. See figure 5-10.

Figure 5-10. Weaver Prone.

Supported Kneeling

If the prone position cannot be used because of the height of the support, the Marine may use the supported kneeling position. The kneeling position allows firing of the pistol from all sides. This position may be altered to maximize the use of cover or support by assuming a variation of the kneeling position (high, medium, low, or two-knee). The kneeling position provides more mobility than the prone position. See figure 5-11.

Caution

The Marine must not indicate to the enemy his position with his exposed knee. If the Marine changes knees while exposed to the enemy, he has indicated the direction of his next shot to the enemy.

Figure 5-11. Supported Kneeling.

Supported Standing

When use of the support is maximized, the supported standing position can be as stable as the supported kneeling or prone position. The supported standing position provides greater mobility than the other positions and usually provides greater observation of the enemy. In the standing position, the Marine must not allow the placement of his foot to indicate to the enemy his position behind cover. See figure 5-12.

Figure 5-12. Supported Standing.

Locate and Engage Targets From Behind Cover

To locate targets when behind cover or to ensure the area is clear before moving, the Marine must expose himself to the enemy. There are two techniques used to locate and engage targets from behind cover: the pie and the rollout. These techniques minimize the Marine's exposure to enemy fire while placing him in a position to engage targets or to move to another location if necessary. Both techniques are used in the kneeling and standing positions.

Pie Technique

The Weaver position is the most effective position when executing the pie technique because the position of the Marine's body minimizes exposure to the enemy. See figure 5-13.

Figure 5-13. Pie Technique.

To perform the pie technique—

- Stay behind cover while moving back and away from the leading edge of the cover. The surroundings and situation dictate the distance the Marine moves back and away from the cover. Generally, the further back the Marine is from cover, the greater his area of observation; staying too close to cover decreases the area of observation.
- Assume a firing position and lower the pistol sights enough to have a clear field of view, aiming in on the leading edge of the cover.
- Take small side steps and slowly move out from behind cover, covering the field of view with the aiming eye and muzzle of the pistol. Wherever the eyes move, the muzzle should move (eyes, muzzle, target).
- Continue moving out from cover until a target is identified or the area is found to be clear. If a target is identified, sweep the safety off, place the finger on the trigger, and engage the target.

Note: If a target is identified before moving out from cover, the pistol should be thumb-cocked and off safe before moving out.

Rollout Technique

In this technique, the Isosceles position is the most effective position when executing the rollout technique because the position of the body allows the Marine to better maintain his balance. See figure 5-14. To perform the rollout technique—

- Stay behind cover, move back, and position the body so it is in line with the leading edge of the cover, ensuring that no part of the body extends beyond cover.
- Assume firing position and come to the Ready, ensuring the muzzle is just behind cover.
- Cant the head and pistol slightly and roll the upper body out to the side enough to obtain a clear field of view. Keep the feet in place and push up on the ball of one foot to facilitate rolling out.
- Continue to roll out from cover until a target is identified or the area is found to be clear. If a target is identified, sweep the safety off, place the finger on the trigger, and engage the target.

Figure 5-14. Rollout Technique.

Note: If a target is identified before moving out from cover, the pistol should be thumb-cocked and off safe before moving out.

Combining the Pie and Rollout Techniques

In some situations, it may be necessary to utilize both the pie and rollout technique in order to search an entire area for targets (i.e., corner of a building, doorway). Changing from one technique to another may allow the Marine to minimize his exposure to the enemy and reduce the time he is exposed to enemy fire.

CHAPTER 6
PRESENTATION OF THE M9 SERVICE PISTOL

In combat, targets present themselves with little or no warning. The Marine must have the ability to react quickly and to effectively and efficiently present the pistol, whether the pistol is in the holster or at a carry. To successfully engage a combat target with the M9 service pistol, the Marine must master pistol presentation from the carries and transport, pistol presentation while assuming a firing position, and search and assess techniques.

Note: The procedures in this chapter are written for right-handed Marines; left-handed Marines should reverse directions as needed.

Sight Alignment/Sight Picture

Pistol presentation drills and dry fire help the Marine achieve a consistent grip and rapid presentation and aid in quickly acquiring sight alignment and sight picture. The Marine always executes the following steps:

- Identify the target and quickly present the pistol to the target while simultaneously sweeping the safety with the thumb of the right hand. (Disengaging the safety with the right thumb ensures the trigger is not pulled before taking the pistol off safe.)
- Shift the focus from the target to the front sight to obtain sight alignment while presenting the pistol. As the front sight breaks the plane of vision, acquire the front sight and begin to apply trigger control as sight picture is acquired.
- Apply trigger pressure until the shot is fired.

Presentation From the Carries and Holsters

Presentation From the Ready

To present the pistol from the Ready, the Marine performs the following steps in sequence. When a target appears—

- Sweep the safety with the thumb of the right hand, place the trigger finger on the trigger, and raise the arms to bring the pistol to the target.

  Note: If the Marine wishes to thumbcock the pistol for a single action shot, the pistol is thumbcocked with the left thumb after the safety is swept with the right thumb. The grip of the left hand may have to be broken to thumbcock the pistol; re-establish the grip after thumbcocking.

- Acquire sight alignment and sight picture within the aiming area and apply trigger pressure until the shot is fired.

Presentation From the Alert

To present the pistol from the Alert, the Marine performs the following steps in sequence. When a target appears—

- Sweep the safety with the thumb of the right hand, place the trigger finger on the trigger, and bring the pistol to the target:
  - If the arms are straight, raise the arms to a 45-degree angle with the deck.
  - If the arms are bent, straighten the arms out toward the target.

Note: If the Marine wishes to thumbcock the pistol for a single action shot, the pistol is thumbcocked with the left thumb after the safety is swept with the right thumb. The grip of the left hand may have to be broken to thumbcock the pistol; re-establish the grip after thumbcocking.

- Acquire sight alignment and sight picture within the aiming area and apply trigger pressure until the shot is fired.

Presentation From a Holster Transport

The pistol is presented from the holster in one continuous, fluid motion.

M12 Holster

To present the pistol from the M12 holster transport, perform the following steps in sequence once the target appears:

- Place the heel of the left hand at the center of the torso with the fingers extended toward the target. (The placement of the left hand allows a two-handed grip to be established in a minimum amount of movement.) At the same time, unfasten and release the D-ring with the right hand.
- Use the right hand to place the thumb on the forward edge of the holster and the fingers around the back edge of the holster, keeping the trigger finger straight.
- Slide the hand up the holster until the fingers come in contact with the pistol grip. At the same time, keep the thumb above the pistol to guide the holster flap up.
- Grasp the pistol grip with the fingers and draw the pistol straight up. Continue withdrawing the pistol while moving the thumb to a position on the safety.

Note: Hand placement should allow the firing grip to be established once the thumb disengages the safety. Any adjustments made to the firing grip after the safety has been disengaged should be minor.

- Once the muzzle clears the holster, rotate the muzzle forward while sweeping the safety.
- Establish a two-handed grip on the pistol by joining the left hand with the right hand in front of the torso. At the same time, start to straighten the pistol out toward the target. See figure 6-1.

Figure 6-1. Presentation From the M12 Holster—Step 1.

- Continue moving the pistol toward the target and, at the same time, place the trigger finger on the trigger, acquire sight alignment and sight picture within the aiming area, and apply trigger pressure until the shot is fired. See figure 6-2.

Figure 6-2. Presentation From the M12 Holster—Step 2.

When time permits to thumbcock the pistol for a single action shot (e.g., a long-range engagement), the Marine presents the pistol from the holster by performing the following steps once the target appears:

- Withdraw the pistol from the holster.
- Sweep the safety with the thumb of the right hand.
- Join the left and right hands and thumbcock the pistol with the left thumb.
- Establish a two-handed grip on the pistol.
- Straighten the pistol out toward the target and, at the same time, place the trigger finger on the trigger, acquire sight alignment and sight picture within the aiming area, and apply trigger pressure until the shot is fired.

M7 Shoulder Holster

The M9 service pistol fits very snugly in the M7 shoulder holster because the holster was originally designed for the M1911A1 .45-caliber pistol, which has a more slim, round design, particularly around the trigger guard. A firm grip is required when holstering and withdrawing the M9 service pistol from the M7 shoulder holster. To present the pistol from the M7 shoulder holster, perform the following steps:

- Unsnap the thumb snap closure with the left hand.
- Wrap the fingers of the right hand around the pistol grip and rest the thumb on top of the inside of the holster. See figure 6-3.

Figure 6-3. Presentation From the M7 Shoulder Holster—Step 1.

- Grasp the pistol grip firmly and draw the pistol up and away from the holster while rotating the thumb in a position to operate the safety.
- Continue withdrawing the pistol until the muzzle clears the holster and rotate the muzzle toward the target. See figure 6-4.

Figure 6-4. Presentation From the M7 Shoulder Holster—Step 2.

- Sweep the safety with the thumb of the right hand while starting to punch the pistol out toward the target.
- Establish a two-handed grip on the pistol by joining the right hand with the left hand in the center of the torso. See figure 6-5.
- Continue punching the pistol out and, at the same time, place the trigger finger on the trigger, establish sight alignment and sight picture within the aiming area, and continue trigger pressure until the shot is fired.

Assault Holster

To present the pistol from the assault holster, perform the following in one, continuous motion:

Figure 6-5. Presentation From the M7 Shoulder Holster—Step 3.

- Bring the heel of the hand down on the hammer of the pistol so that it disengages the thumb break. See figure 6-6.
- Sweep the hand forward in a small circular motion, slide the hand up the holster, and bring the fingers up under the pistol grip.

Note: Depending on the type of retention strap, this circular motion releases the retention strap or pushes the retention strap forward and out of the way.

Figure 6-6. Presentation From the Assault Holster—Step 1.

- Grasp the pistol grip and draw the pistol straight up and out of the holster while establishing a firing grip and positioning the thumb on the safety to operate it. See figure 6-7.

Figure 6-7. Presentation From the Assault Holster—Step 2.

Concealed Pistol Holster

Depending on the clothing worn by the Marine, there are two methods for accessing the concealed pistol holster so that the clothing does not become an obstruction while presenting the pistol. To present the pistol from the concealed pistol holster, the following steps are performed in one, continuous motion when the target appears:

- If the Marine is wearing a short jacket/sweater (waist-level) that is buttoned or zipped—
  - Move the left hand across the body and grasp the jacket or sweater just above the holster, at the same time place the right hand on the holster below the jacket or sweater. See figure 6-8 on page 6-6.
  - Pull up on the jacket or sweater with the left hand. At the same time, slide the right hand up the holster until the fingers come in contact with the pistol grip. Keep the thumb above the pistol to ensure the jacket or sweater clears the pistol. See figure 6-9 on page 6-6.

Note: Leaning slightly forward at the waist may assist in clearing the jacket or sweater from the pistol and in removing the pistol from the holster.

- If the Marine is wearing a long jacket (below waist-level), or a short jacket or sweater that is unfastened—
  - Place the heel of the left hand at the center of the torso to hold the left side of the jacket down with the fingers extended. At the same time, place the right hand on the right side of the jacket, with the thumb underneath the right side of the jacket below the left hand. See figure 6-10 on page 6-7.
  - Keep thumb of the right hand against the body and throw the jacket back and away from the holster. See figure 6-11 on page 6-7.

Note: Keeping a weighted object (e.g., keys, extra magazine) in the right-hand pocket of the jacket assists in throwing the jacket back and away from the holster.

  - Place the right hand on the holster and slide the hand up the holster until the fingers come in contact with the pistol grip.
- Grasp the pistol grip with the fingers and draw the pistol straight up. Continue withdrawing the pistol while moving the thumb to a position on the safety.

Note: Hand placement should allow the firing grip to be established once the thumb disengages the safety. Any adjustments made to the firing grip after the safety is disengaged should be minor.

- Rotate the muzzle forward, once it clears the holster, while sweeping the safety.
- Establish a two-handed grip on the pistol by joining the left hand with the right hand in front of the torso. At the same time, start to punch the pistol out toward the target.

Figure 6-8. Presentation From the Concealed Pistol Holster—Step 1.

Figure 6-9. Presentation From the Concealed Pistol Holster—Step 2.

Figure 6-10. Presentation From the Concealed Pistol Holster with the Long Jacket—Step 1.

Figure 6-11. Presentation From the Concealed Pistol Holster with the Long Jacket—Step 2.

- Continue punching the pistol out and, at the same time, place the trigger finger on the trigger, acquire sight alignment and sight picture within the aiming area, and apply trigger pressure until the shot is fired.

Presentation While Assuming the Kneeling Position

In combat, the Marine must be able to quickly assume the kneeling position while presenting the pistol to the target. This usually takes advantage of cover and provides a more stable base for shooting. The following procedures are based on the Isosceles and Weaver variations of the standing to the kneeling shooting positions.

From the Ready

The Marine drops to the kneeling position while raising the arms to bring the pistol up to the target.

> *Note:* The Weaver variation maintains the body's position in a 40 to 60 degree oblique to the target.

At the same time, the Marine sweeps the safety with the thumb of the right hand, places the trigger finger on the trigger, acquires sight alignment and sight picture within the aiming area, and applies trigger pressure until the shot is fired.

From the Alert (Arms Straight)

The Marine drops to the kneeling position while raising the arms to bring the pistol up to the target.

> *Note:* The Weaver variation maintains the body's position in a 40 to 60 degree oblique to the target.

At the same time, the Marine sweeps the safety with the thumb of the right hand, places the trigger finger on the trigger, acquires sight alignment and sight picture within the aiming area, and applies trigger pressure until the shot is fired.

From the Alert (Close Quarters, Elbows Bent)

The Marine drops to the kneeling position while pushing the arms out toward the target.

> *Note:* The Weaver variation maintains the body's position in a 40 to 60 degree oblique to the target.

At the same time, the Marine sweeps the safety with the thumb of the right hand, places the trigger finger on the trigger, acquires sight alignment and sight picture within the aiming area, and applies trigger pressure until the shot is fired.

From the Holster Transport

The Marine withdraws the pistol from the holster while dropping to the kneeling position. The pistol should be rotated to the target by the time the knee hits the deck. The Marine sweeps the safety with the thumb of the right hand, establishes a two-handed firing grip and—

- In the Weaver variation, raises the arms to bring the pistol up toward the target.
- In the Isosceles variation, pushes the pistol out toward the target.

At the same time, the Marine places the trigger finger on the trigger, acquires sight alignment and sight picture within the aiming area, and applies trigger pressure until the shot is fired.

Presentation While Assuming the Prone Position

The prone position is generally assumed to take advantage of cover or to provide additional stability for shooting. The following procedures are executed from the standing holster transports.

Isosceles Prone

The Marine withdraws the pistol from holster. At the same time, the Marine eliminates body contact with the deck by either the squat method or the drop method. The pistol is rotated to the target as the left hand is placed on the deck. The Marine sweeps the safety off.

> *Note:* If the Marine wishes to thumbcock the pistol for a single action shot, it is done once the Marine is on the deck.

The Marine establishes a two-handed firing grip while spreading the legs a comfortable distance apart for stability. He places the trigger finger on the trigger, acquires sight alignment and sight picture within the aiming area, and applies trigger pressure until the shot is fired.

Weaver Prone

The Marine withdraws the pistol from the holster. At the same time, the Marine eliminates body contact with the deck by either the squat method or the drop method. The Marine maintains the 40 to 60 degree oblique to the target. The pistol should be rotated to the target as the left hand is placed on the deck. The Marine sweeps the safety off.

> *Note:* If the Marine wishes to thumbcock the pistol for a single action shot, it is done once the Marine is on the deck.

The Marine brings the left knee up, establishes a two-handed firing grip, and places the left elbow on the deck for stability. He places the trigger finger on the trigger, acquires sight alignment and sight picture within the aiming area, and applies trigger pressure until the shot is fired.

Search and Assess

To be successful in combat, the Marine must have the ability to assess the situation and take appropriate action following engagement. Once the Marine fires one or two rounds to engage a target, he must quickly assess the situation and the effectiveness of his engagement. This split-second assessment allows the Marine to determine the best course of action and prevents him from expending a number of rounds without assessing the results and his ability to accurately engage the target.

To search and assess, perform the following steps in sequence:

- Place the trigger finger straight along the receiver immediately after a target is engaged. Do not place the pistol on safe.
- Lower the arms just enough to look over the pistol sights and provide a clear field of view. Do not rotate the wrists in order to angle the pistol downward, this breaks the firing grip and changes the tension in the arms. Lowering the arms maintains the firing grip, keeps the sights level, and allows sight alignment and sight picture to be quickly re-established should follow-on shots need to be fired.
- Search the area by moving the head, eyes, and pistol left and right (approximately 45 degrees from center). Wherever the head moves, the muzzle moves (eyes, muzzle, target). Keeping both eyes open increases the field of view.

If the target has not been eliminated, the Marine must determine whether to re-engage the target:

- If the Marine decides to re-engage the target, the target's size, time and distance to the target, and Marine's capabilities dictate the technique used. Re-engagement techniques include re-engaging the target with a well-aimed precision shot, two shots, or offset aiming.
- If the target's distance exceeds the Marine's engagement capabilities or if the target is partially exposed and too small for the Marine to accurately engage, the Marine should not attempt to re-engage the target. The Marine may seek cover, seek out a better opportunity for engagement, or use support to better stabilize the pistol.

When it is determined that the area is clear of all enemy threat, place the pistol on safe without breaking the grip and assume a pistol carry or pistol transport.

CHAPTER 7
PISTOL ENGAGEMENT TECHNIQUES

The Marine must be able to quickly detect targets and employ the proper pistol engagement and firing techniques. The Marine must also be able to re-engage a target if the initial engagement is not successful.

Note: The procedures in this chapter are written for right-handed Marines; left-handed Marines should reverse directions as needed.

Target Detection

Target indicators reveal an enemy's position. Most combat targets are detected by smoke, flash, dust, noise, or movement and are usually only momentarily visible. Target indicators are grouped into three general areas: movement, sound, and improper camouflage.

Movement

The Marine does not need to look directly at an object to notice movement. The eye is attracted to any movement, especially sudden movement. A slowly moving target is harder to detect than one with quick, jerky movements. Therefore, the ability to locate a moving target depends primarily on the speed of the object's movement.

Sound

Sound (e.g., movement, rattling equipment, talking) can be used to detect an enemy position. Sound provides only a general location, making it difficult to pinpoint a target by sound alone. However, sound can alert the Marine to the presence of a target and increases the probability of locating the target through other indicators.

Improper Camouflage

The improper use of camouflage creates three target indicators: shine, outline, and contrast with the background. The closeness of a typical pistol engagement somewhat diminishes these effects, but can still affect target detection. The Marine's ability to recognize target indicators aids in the detection of targets.

Shine

Shine is created from reflective objects such as metal, glass, pools of water, and the skin's natural oils. Shine acts as a beacon in locating a target's position.

Outline

The outline of camouflaged objects such as the body, head and shoulders, weapons, and gear can be recognizable.

Contrast With the Background

Objects contrast with a background because of differences in color, surface, and shape. The following are examples of objects contrasting with their background:

- A target wearing a dark uniform would be clearly visible in an area of snow or sand.
- Symmetrical shapes, such as helmets or rifle barrels, can be detected in a wooded area.
- Fresh soil around a fighting hole contrasts with the otherwise unbroken ground surface.

While observing an area, the Marine notes anything that looks out of place or unusual and studies it in more detail in order to increase the chances of spotting a hidden threat.

Techniques of Fire

To successfully engage a combat target with the M9 service pistol, the Marine must employ effective techniques of fire. The Marine's performance of these skills and the proper application of the fundamentals of marksmanship are critical to success in a combat situation.

Double and Single Action Firing

When the M9 service pistol is taken off safe, it is capable of firing in both a double and a single action mode.

Double Action Mode

The design of the M9 service pistol causes the first shot fired to be a double action shot.

In double action firing, two actions occur as the trigger is moved to the rear; the hammer moves to the rear, cocking the pistol, and then the hammer moves forward, firing the pistol. More pressure is required on the trigger to fire a double action shot due to the distance between the trigger and hammer and the weight of the trigger. A double action shot requires approximately 9 to 16 pounds of pressure to move the trigger rearward.

Maintaining sight alignment and sight picture are harder when firing a double action shot. Therefore, the sights will more than likely move outside the aiming area when applying trigger pressure.

Single Action Mode

In single action firing, the pistol is already cocked because the cycle of operation in the first double action shot leaves the hammer cocked to the rear. Therefore, the only action taking place as the trigger is moved to the rear is the hammer moving forward, firing the pistol. A single action shot requires approximately 4 to 6 pounds of pressure to move the trigger rearward. Therefore, the application of trigger control is easier when firing a single action shot.

To enable the first shot to be fired single action, the pistol's hammer can be manually cocked with the thumb:

> *Note:* The pistol must be taken off safe before it can be thumbcocked.

- Use the left thumb to pull back on the hammer to cock it. This ensures that the firing grip of the right hand does not have to be broken.
- Ensure that the hammer moves all the way to the rear.
- Ensure the trigger finger remains straight along the receiver until the pistol is fully cocked.
- Re-establish the firing grip with both hands once the pistol is cocked.

Single Action or Double Action Firing Factors

The decision to fire in a single or double action mode is made rapidly. Ultimately, the decision is based on the Marine's abilities, but the decision is also based on time and accuracy. The pistol is fired in the double action mode when trigger control, sight picture, and stability of hold are not as critical for accuracy (i.e., close range, large targets). The pistol is fired in the single action mode when the fundamentals are more critical to accuracy (i.e., long range, small targets).

Time

The Marine sacrifices time to fire a single action, precision shot; but what he sacrifices in time, he gains in accuracy.

For quick engagements at close range, there may not be time to thumbcock the pistol for a single action shot. Therefore, firing the first shot in the double action mode is preferred because shots are needed on target quickly and stability of hold and sight picture are not as critical to accuracy.

When time permits and for targets at longer ranges, the pistol may be thumbcocked to place it in the single action mode to reduce the weight of the trigger and the distance the trigger must travel rearward to fire the first shot.

Distance and Size of the Target

The smaller the target, the more critical the application of the fundamentals in order to engage the target accurately. To accurately engage a small target (e.g., head shot, long-range target) it is better to thumbcock the pistol for a single action shot. Engagement of a smaller target requires additional precision because sight alignment and sight picture are more critical to accuracy.

Two-Shot Technique

During combat, the preferred pistol engagement technique is to rapidly fire more than one shot on a target to eliminate it as a threat. Two shots fired in rapid succession increase the trauma (i.e., shock, blood loss) on the target and increase the Marine's chances of quickly eliminating the threat. Therefore, two shots are most often fired in rapid succession on a target at close range.

After the pistol is fired, the muzzle climbs with the recoil of the pistol. To fire two shots, the Marine must quickly recover the sights to the same area on the target while re-acquiring sight alignment and sight picture. The proper recovery automatically brings the sights back on target following recoil. A quick recovery allows more time for the Marine to align the sights and apply trigger control to fire the next shot. Recovery begins immediately after the application of the fundamentals to bring the pistol sights into alignment with the target in preparation for firing the next shot.

The amount the muzzle climbs during recoil depends on the amount of controlled muscular tension in the grip and wrists used to stabilize the pistol and create consistency in resistance to recoil. Controlled muscular tension allows the pistol sights to recover consistently back on target within a minimum amount of time. The speed of delivery of multiple shots depends on how fast the Marine can re-acquire sight alignment. Sight recovery is determined by the following:

- The key to proper recovery is a stable firing position and proper grip.
- If the Marine's firing position is not stable, recoil forces him out of his firing position, requiring him to re-establish his position before he takes his next shot.
- An improper grip or lack of controlled muscular tension causes the pistol to move in the Marine's hand after the shot is fired, disrupting sight alignment and requiring the Marine to re-establish his grip.

Slow Fire Technique

Sight alignment becomes more critical the smaller the target and the greater the distance to the target. In these situations, the Marine does not engage the target with two rapidly fired shots because he has to slow down his application of the fundamentals in order to fire a precision shot(s). To engage small targets (i.e., partially exposed) and targets at longer ranges where precision is required, the Marine must employ the following slow fire technique:

- Thumbcock the pistol for a single action shot.
- Slow down application of the fundamentals.
- Fire one well-aimed, precision shot on target.

Re-engagement Techniques

Once the Marine has quickly assessed the situation and determined that the threat still exists, he may make the decision to re-engage the target to eliminate it as a threat. Re-engagement techniques include engaging the target with a precision shot through slow fire, two shots, or offset aiming. The time, size, and distance to the target as well as the Marine's marksmanship abilities dictate the technique used to re-engage the target.

Slow Fire Technique

If the Marine has engaged a target and the target still poses a threat, the Marine may choose to slow down his application of the fundamentals and fire a slow fire, precision shot. The placement of one well-aimed, precision shot on a designated area of the target increases the chances of eliminating the target as a threat. However, the time the Marine has to engage the target determines if he can slow down his application of the fundamentals to fire a precision shot. Firing a precision shot takes time, so the Marine will sacrifice time for accuracy.

The key to successful shot placement is the Marine's ability to slow down his performance and focus on the application of the fundamentals of marksmanship. The Marine must determine where to place shot for maximum effectiveness. In order to make this decision, the Marine must consider distance and size of the target and time. For targets at close range and within the Marine's marksmanship abilities, a precision shot is placed in the head to immediately eliminate the target as a threat. Sometimes, even at close ranges, the target may only be partially exposed; therefore, the Marine is presented with a small target and aims his sights on the portion of the target that is exposed. For targets at long ranges, the target is smaller and requires a precision shot. A precision shot may be placed in the body to add trauma to the target and increase the chances of eliminating the target as a threat.

Two-Shot Technique

If a target is accurately engaged, but does not go down, the Marine may re-engage the target with additional shots. Two additional, rapidly delivered shots increases the trauma and increases the Marine's chances of eliminating the threat.

The size and distance to the target affect how quickly two shots can be delivered on the target. The speed at which two shots are fired also depends on the Marine's marksmanship abilities and how fast he can re-acquire his front sight. However, the Marine must not compromise accuracy for speed; the key to successful target engagement is to fire only as quickly as the Marine can fire effectively.

Offset Aiming

When the Marine assesses the situation and determines his shots are not successful and not striking the target in the designated aiming area, he may employ offset aiming. Since the pistol's sights cannot be adjusted, offset aiming is applied to adjust the aiming area and to cause rounds to strike center mass. This technique should only be applied when the Marine determines that he is applying the fundamentals of marksmanship correctly. Sight picture will change as the aiming area is adjusted.

The known strike of the round offset aiming technique requires shifting the point of aim to compensate for rounds striking off target center. To effectively engage a target using this technique, the Marine must be able to see where the rounds are striking and then aim an equal distance from the center of the aiming area opposite the observed strike of the round. For example, if the rounds are striking the target high and left of center mass, aim an equal and opposite distance low and right.

Multiple Targets

If engaging more than one target at a time, the Marine adheres to the fundamentals of marksmanship and employs the techniques of multiple target engagement. The introduction of multiple targets in a combat scenario requires additional skills that must be learned and practiced if a Marine is to be successful. To be effective in combat, the Marine must be able to detect targets through identification of target indicators (identified on page 7-1), prioritize the targets, and employ multiple target engagement techniques.

Combat Mindset

Successful engagement of multiple targets requires a somewhat different mindset from single target engagement. For example, following engagement of a single target, the Marine assesses the situation. During multiple target engagements, after the first target is engaged, he must immediately engage the next target and continue until all targets have been eliminated. Because split-second decisions must be made, the development of a combat mindset is important to success on the battlefield. A combat mindset allows the Marine to control the pace of the battle rather than react to the threat.

Mental preparedness is essential to successful engagement of multiple targets, and the required mindset must be developed until it becomes second nature to the Marine. When multiple targets appear, the Marine must prioritize the targets to establish an engagement sequence. To prioritize targets and establish an engagement sequence, the Marine must be aware of the surroundings, not focus on one target, and continuously search the terrain for additional targets.

Prioritizing Targets

Once multiple targets have been identified, they must be prioritized in terms of the threat each target presents. While the fundamentals of marksmanship must still be applied, prioritizing targets and planning the engagement are just as essential to successful multiple target engagement.

Target priority is based on factors such as proximity, threat, and opportunity. It also encompasses the Marine's proficiency level since a Marine should not attempt to engage a target beyond his proficiency level or the pistol's capability (e.g., a target 100 yards away). The principal method of prioritizing targets is to determine the level of threat for each target so that all targets may be engaged in succession, starting with the most threatening and ending with the least threatening.

Prioritizing targets is an ongoing process. As the engagement proceeds, new targets may appear or previously identified targets may take cover, delaying their engagement. Therefore, the Marine must remain constantly alert to changes in target threat, proximity, and the opportunity for engagement and revise target priorities accordingly.

Engaging Multiple Targets

Once targets have been prioritized, the Marine must quickly eliminate them. It is crucial to understand and practice the physical techniques for engaging multiple targets until they become second nature. When these techniques occur as automatic responses, the Marine maintains an awareness of the complete battlefield and is able to concentrate on the mental aspects of multiple target engagement.

Engagement Methods

The fundamentals of marksmanship are critical to the development of skills to support multiple target engagement. While responsiveness is important when engaging multiple targets, the Marine's primary concern is to place accurate rounds on target. A balance must be struck between placing shots quickly on targets at close range and slowing down to place precision shots on targets at long range because all targets pose a threat and must be accurately engaged. To engage multiple targets—

- Engage every target, move from target to target.
- After all of the targets have been initially engaged, assess the effectiveness of the engagements and, if necessary, re-engage targets that still pose a threat.
- If all of the targets are eliminated, search the area for new targets.

Considerations for Field Firing Positions

The selection and use of field firing positions are critical to engagement of multiple targets. As in any firing situation, if the situation permits, the Marine should make a quick mental review of the terrain to select a firing position that provides stability, mobility, and observation of the enemy. The firing position should also provide flexibility for engaging multiple targets. The more dispersed the multiple targets, the greater the lateral movement the position must afford to engage them.

The prone position provides the maximum stability for firing well-placed shots, but engaging multiple targets from this position may require adjustment in the position from shot to shot depending on the distance between the targets and their location. Because the arms are fully extended on the ground, the prone position can be restrictive, increasing the time it will take to recover the sights onto subsequent targets. If the Marine must make an adjustment in the prone position to engage multiple targets, he must ensure that he maintains proper arm placement. Improper arm placement may affect his stability of hold and his ability to acquire sight picture. The following identifies position adjustments:

- For targets that are not widely dispersed, the Marine adjusts the position of the lower part of the body to orient the upper body toward the target without disturbing the placement of the arms. This allows the Marine to maintain stability of hold and quickly recover on subsequent targets. Moving the legs to the left orients the upper body to the right; moving the legs to the right orients the upper body to the left.
- For targets that are widely dispersed, the Marine may need to adjust his entire position to orient his body in the direction of subsequent targets. Using the left hand, push the upper body off the deck just high enough to move the body in the direction of the target. Extend the right arm toward the target and keep the muzzle pointed in the direction of the target.

The kneeling and standing positions provide an increased field of view and allow maximum lateral adjustment to engage dispersed targets. The type of adjustment the Marine must make in the kneeling or standing position in order to orient his body in the direction of each target is related to the distance between the targets. It is also related to the Marine's ability to maintain stability of hold and recover from recoil. Once the first target is engaged—

- Maintain a proper shooting position, rotate the body in the direction of the target, keeping the feet in place. In the standing position, the Marine may bend slightly at the knees while rotating the body. This allows the Marine to distribute his weight forward to provide additional stability to the position and to better control the effects of recoil.
- Adjust the entire position by either adjusting the placement of the feet or knee(s) in order to face in the direction of the new target if the targets are some distance apart.
- Ensure that as the eyes move to a new target, the head and pistol muzzle should follow.

Moving Targets

The majority of combat targets will be moving; the enemy will move quickly from cover to cover, exposing himself for the shortest time possible. Therefore, a moving target must be engaged before it disappears. The Marine must engage a moving target with the same accuracy and precision used to engage a stationary target. Practice in the proper application of leads and the fundamentals of marksmanship enable the successful engagement of moving targets.

Types of Moving Targets

Moving Target

This type of target moves in a consistent manner (i.e., walking or running) and is in continuous sight as it moves across the Marine's field of vision. However, unless the enemy is completely unaware of the Marine's presence, this type of target is not likely to present itself.

Stop and Go Target

This type of target appears and disappears during its movement due to intermittent cover. It will present itself for only a short period of time before re-establishing cover. This type of target is most vulnerable to fire at the beginning and end of its movement because it begins slow and gains momentum as it leaves cover and then slows to avoid overrunning the new cover.

Leads for Moving Targets

Once the moving target has been identified it must be engaged. To engage a moving target, the Marine must aim at a point in front of the target, which is known as a lead. Lead is the distance in advance of the target that is required to strike the target when it is moving. To achieve success, determining the amount of lead to engage a moving target must be as precise as possible. When a shot is fired at a moving target, the target continues to move during the time of the bullet's flight. For this reason, the aim must be in front of the target; otherwise, the shot will fall behind the target.

Factors Affecting Lead

Factors that affect the amount of lead are the target's range, speed, and angle of movement.

Range. There is a time lag from the time a round is fired until the round strikes at the point of aim. This time of flight could allow a target to move out of the bullet's path if the round were fired directly at the target. Time of flight increases as range to the target increases. Therefore, the lead must be increased in proportion to the distance to the target.

Speed. A greater lead is required to hit a running man than a walking man because the running man moves a greater distance during the flight of the bullet.

Angle of Movement. The angle of target movement also affects the amount of lead required for target engagement. The angle of movement across the Marine's line of sight relative to the flight of the bullet determines the type (amount) of lead.

Types of Leads

Full Lead. The target moves straight across the Marine's line of sight with only one arm and half the body visible. This target requires a full lead because it moves the greatest distance across the Marine's line of sight during the bullet's flight.

Half Lead. The target moves obliquely across the Marine's line of sight (at about a 45-degree angle). One arm and over half of the back or chest are visible. This target requires half of a full lead because it moves half as far as a target moving directly across the Marine's line of sight during the bullet's flight.

No Lead. A target moving directly toward or away from the Marine presents a full view of both arms and the entire back or chest. No lead is required. This target is engaged in the same manner as a stationary target because it is not moving across the Marine's line of sight.

Point of Aim Technique

To engage a moving target, a Marine establishes a lead using a point of aim offset aiming technique. This technique uses predetermined points of aim to sector a man-sized target vertically, halfway between center mass and the leading edge of the target (both one point of aim and two points of aim) (see fig. 7-1 on page 7-8). The same units of measure can be applied off the target for holds of additional points of aim. To use the point of aim technique to establish a lead on a man-sized moving target at various ranges, speeds, and angles of movement, the following guidelines apply. See figure 7-2 on page 7-9.

- For a target moving at a distance of 15 yards away or less, no lead is required.
- For a target walking directly across the Marines line of sight (full lead) between 16 and 25 yards away, hold 1 point of aim in the direction the target is moving.
- For a target running directly across the Marine's line of sight (full lead) between 16 and 25 yards away, hold 2 points of aim in the direction the target is moving.

Methods of Engagement

Moving targets, although difficult, can be engaged by the tracking or ambush method or a combination of the two. See figure 7-2 on page 7-9.

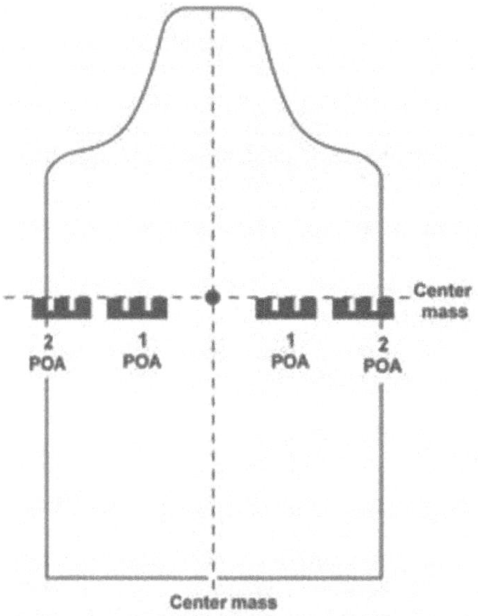

Figure 7-1. Points of Aim.

Tracking Method

In this method, the Marine "tracks" or follows the target with his front sight while maintaining sight alignment and a point of aim on or ahead of (leading) the target until the shot is fired. Sight picture is the aiming point in relation to the target while maintaining sight alignment (when a lead is established in moving target engagement, the sights are not entered on the target). To execute the tracking method—

- Thumbcock the pistol while presenting it to the target.
- Track the muzzle of the pistol through the target to the desired point of aim (lead). The point of aim may be on the target or some point in front of the target depending upon the target's range, speed, and angle of movement.
- Track and maintain focus on the front sight while applying trigger pressure and acquiring sight alignment.
- Continue tracking and applying trigger pressure while acquiring sight picture. When sight picture is established, engage the target while maintaining the proper point of aim (lead).
- Follow through so the lead is maintained as the bullet exits the muzzle. Continuing to track also enables a second shot to be fired on target.

Ambush Method

The ambush method is generally used to engage a stop and go target and when it is difficult to track the target with the pistol, such as in the prone position. With this method, the pistol is aimed at a predetermined engagement point ahead of the target and along its path, allowed to remain stationary, and fired when the target reaches the predetermined engagement point. The engagement point is based on the required point of aim (lead) to engage the target effectively. When the sights are settled, the target moves into the predetermined engagement point creating the desired sight picture. Once sight picture is established, the remaining pressure is applied on the trigger until the shot is fired. To execute the ambush method—

- Look for a pattern of exposure.
- Thumbcock the pistol while presenting it to a selected point of aim ahead of the target.
- Apply trigger pressure while obtaining sight alignment in the aiming area.
- Continue applying trigger pressure and hold sight alignment until the target moves into the predetermined engagement point and the desired sight picture is established.
- Engage the target once the sight picture is acquired.
- Follow through so the pistol sights are not disturbed as the bullet exits the muzzle.

Applying the Fundamentals of Marksmanship

The engagement of moving targets is a perishable skill that must be practiced frequently if it is to be maintained. The Marine must practice to develop the skill to calculate the point of aim (lead) and

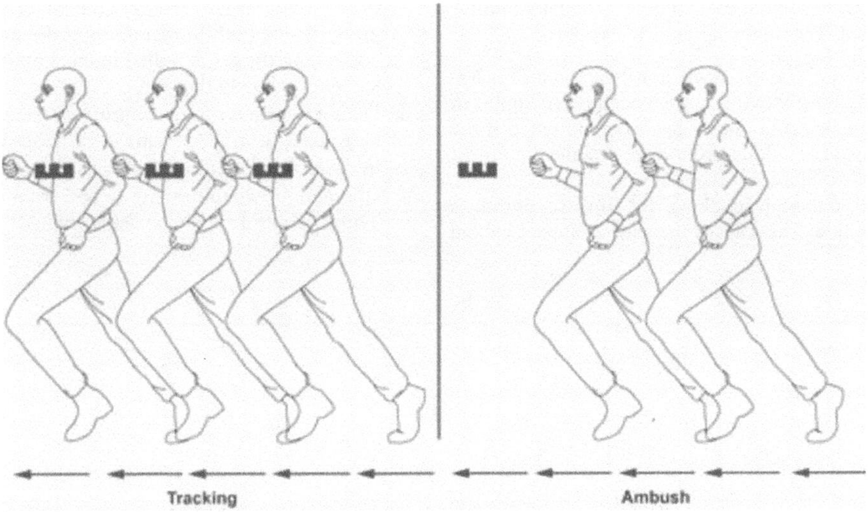

Figure 7-2. Moving Target Engagement Methods.

fire the shot while maintaining the proper point of aim (lead).

To engage moving targets using the tracking method, the Marine moves the pistol smoothly and steadily as the target moves. A stable position and firm grip are necessary to steady the pistol sights while tracking.

When using the tracking method, continue tracking the target while following through with the shot process so the point of aim (lead) is maintained as the bullet exits the muzzle. Continuing to track also enables a second shot to be fired on target if necessary. Concentration should be on continuing to track while applying the fundamentals. The fundamentals must be instinctively applied, allowing concentration on tracking the target and applying point of aim (lead).

Aiming

Sight alignment remains unchanged for accurate engagement of a moving target. The most common error when engaging moving targets is to focus on the target rather than focusing on the front sight. Sight picture is the point of aim in relation to the target while maintaining sight alignment. For both the ambush and tracking methods, sight picture is changed from the normal center mass picture, based on range, speed, and angle of movement of the target. It takes practice in moving target engagement to quickly establish the desired sight picture for a given point of aim (lead).

Breath Control

There is no difference in breath control when engaging moving targets; the breath is held to fire the shot.

Trigger Control

Trigger control is critical to firing shots while establishing and maintaining sight alignment and sight picture. Because the single action mode is the preferred method of engaging a moving target, the pistol is thumbcocked before trigger control is applied. The following guidelines apply:

- Do not use interrupted trigger control when engaging moving targets because the point of aim (lead) is lost or has to be adjusted if the trigger is held in order to re-assume the proper sight picture.
- Do not stop tracking as trigger control is applied. This causes the shot to impact behind the moving target. Trigger control should be uninterrupted while maintaining the point of aim so the trigger is pulled in one continuous, smooth motion to the rear.
- Do not interrupt trigger control when the target is masked from view. This is particularly true in the ambush method for targets that appear to stop and go when moving to and from cover. Trigger control should be uninterrupted.

CHAPTER 8
ONE-HANDED TECHNIQUES

In combat, a situation may arise where the Marine must engage a target by using only one hand. A Marine fires the pistol one-handed when—

- The Marine's other hand is occupied.
- A target presents itself so quickly and at a close range that the Marine does not have time to establish a two-handed grip and position.
- One of the Marine's hands is injured.

Note: The procedures in this chapter are written for right-handed Marines; left-handed Marines should reverse directions as needed.

Presentation and Shooting

Adjustment of the Standing Position

To fire accurately, the Marine must maintain the same stability and control of the pistol, manage recoil, and recover on target as effectively with one hand as with two. This is accomplished by adjusting the standing position as the pistol is presented to the target (see fig. 8-1).

The Marine's angle to the target determines what adjustments are made to the standing position. This adjustment is made while the pistol is being presented to the target by moving or sliding the feet to orient the body to the target. For stability and balance, the right foot may be slightly forward of the left foot. The further to the right of the Marine the target is the farther forward the right foot will be. To adjust the standing position when firing one-handed—

- Increase the muscular tension in the right arm and bend the elbow slightly to better manage the pistol.

Note: For some Marines, muscular tension in the firing arm and the position of the body in relation to the target causes them to cant the pistol inboard slightly. A slight inboard cant of the pistol and firing arm allows a natural bend in the elbow and allows a better management of recoil and recovery of the sights back on target. Any cant should be natural and not caused by excessive muscular tension. A slight cant does not affect shooting performance and should not alter the application of the fundamentals of marksmanship.

- Apply controlled muscular tension throughout the body to better manage the effects of recoil.
- Lean forward aggressively, roll the shoulders forward, and tuck the head into the shoulders.

Figure 8-1. One-Handed Presentation.

Note: The following step is done only during training to demonstrate how to stabilize the position and to simulate the Marine's other hand being occupied.

- Place the left hand in a fist on the center of the torso to increase stability. If the left arm is swinging freely it causes a corresponding movement in the pistol, reducing stability and control. Placing the left hand on the torso also ensures it is not covered by the muzzle of the pistol as the pistol is presented.

Searching and Assessing After Firing

Following one-handed target engagement, the Marine must quickly decide if it is to his advantage and if it is possible to free his occupied hand in order to place both hands on the pistol to search and assess. Whenever possible, the Marine places both hands on the pistol to increase stability and to be ready for re-engagement if necessary.

Reloading

Dry Reload

When only one hand is available, the dry reload can be performed as follows:

- Seek cover, if the situation permits.
- Press the magazine release button and allow the magazine to fall to the deck. If using the right hand, press the magazine release button with the thumb. If using the left hand, press the magazine release button with the trigger finger.
- Point the muzzle in a safe direction and position the pistol to facilitate loading the magazine. Depending on the firing position, this is accomplished using one of the following methods:

  Two-Knee Kneeling Position

  - Rotate the pistol so the magazine well faces up and the muzzle of the pistol is pointed away from the body.
  - Place the pistol between the thighs or knees and apply pressure to hold the pistol in place.
  - Withdraw a filled magazine from the ammunition pocket and insert it into the magazine well, seating it with the heel of the hand. See figure 8-2.

Figure 8-2. One-Handed Dry Reload: Two-Knee Kneeling Position.

Standing Position

- Rotate the pistol so the magazine well faces up and the muzzle of the pistol is pointed away from the body.
- Place the pistol between the thighs or knees and apply pressure to hold the pistol in place. Bend slightly at the knees to secure the pistol between the thighs.
- Withdraw a filled magazine from the ammunition pocket and insert it into the magazine well, seating it with the heel of the hand. See figure 8-3.

Kneeling Position (High, Medium, Low)

- Rotate the pistol so the magazine well faces outboard and the muzzle is pointed away from the body.
- Place the rear portion of the slide in the bend of the right knee and apply pressure with the leg to hold the pistol in place. For additional control, the Marine may drop to a two-knee kneeling position and place the pistol between his knees.
- Withdraw a filled magazine from the ammunition pocket and insert it into the magazine well, seating it with the heel of the hand. See figure 8-4 on page 8-4.

Prone Position

- The Marine places the pistol on the deck in front of him and against his body with the muzzle pointed in a safe direction. Alternately, the Marine may roll on his side and place the pistol between his knees, with the magazine well facing out.
- Withdraw a filled magazine from the ammunition pocket and insert it into the magazine well, seating it with the heel of the hand. See figure 8-5 on page 8-4.

- Grasp the pistol grip and remove the pistol from its secured location or from the deck.
- Press the slide release to allow the slide to move forward and chamber a round. If using the right hand, press the slide release with the thumb. If using the left hand, press the slide release with the index finger.

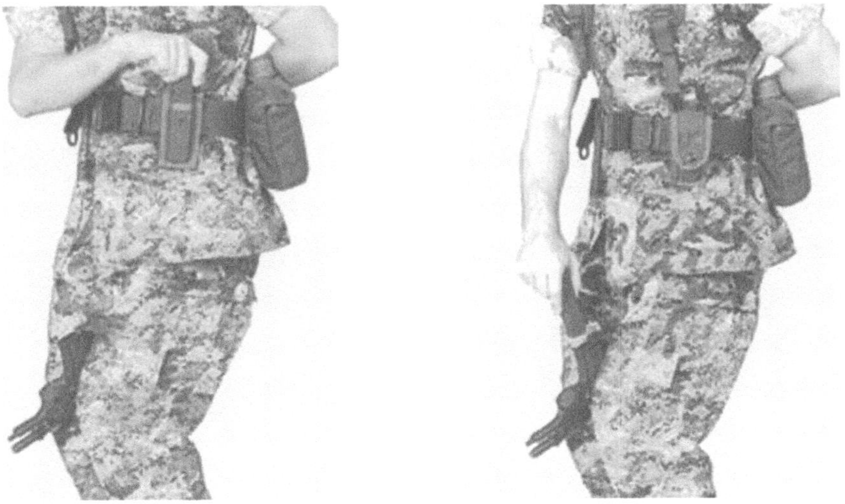

Figure 8-3. One-Handed Dry Reload: Standing Position.

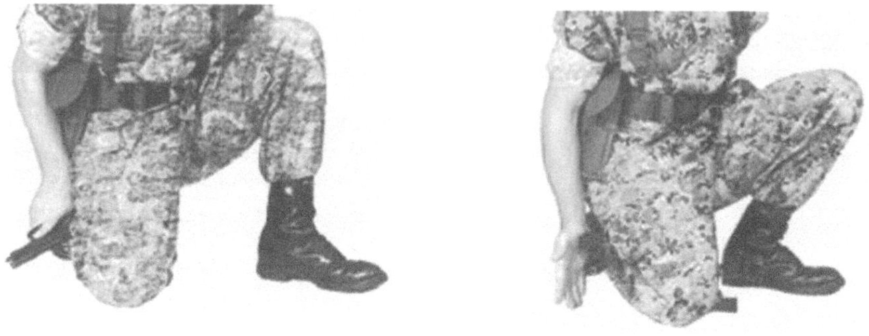

Figure 8-4. One-Handed Dry Reload: Kneeling Position.

Condition 1 Reload

When only one hand is available, a Condition 1 reload can be performed as follows:

- Point the muzzle in a safe direction and position the pistol to allow for removal of the magazine (e.g., between the thighs, in the bend of the knee, in the holster, tucked in the cartridge belt, on the deck against the body). Any position is correct if it allows the magazine well to be exposed.
- Withdraw a filled magazine from the ammunition pocket.
- Press the magazine release button and remove the magazine from the pistol. If using the right hand, press the magazine release button with the thumb. If using the left hand, press the magazine release button with the index finger.
- Insert the filled magazine into the magazine well, seating it with the heel of the hand.
- Stow the partially-filled magazine (e.g., inside the cartridge belt, in the cargo pocket).
- Grasp the pistol grip and remove the pistol from its secured location, keeping the trigger finger straight and off the trigger until ready to fire.

Remedial Action

Remedial action requires investigating the cause of the stoppage, clearing the stoppage, and returning the pistol to operation. When performing remedial action, seek cover if the tactical situation permits. Once a pistol ceases to fire, the Marine must visually or physically observe the pistol to identify the problem before it can be

Figure 8-5. One-Handed Dry Reload: Prone Position.

cleared. The steps taken to clear the pistol are based on what is observed:

- Remove the finger from the trigger and place it straight along the receiver.
- Lock the slide to the rear. To pull and lock the slide to the rear, push up on the slide stop and maintain pressure with the right thumb. (Push up on the slide stop with the left index finger if remedial action is being performed with the weak hand.)
- Secure the rear sight on the top of the cartridge belt or any other surface that provides the resistance needed to stabilize the pistol. See figure 8-6. While applying pressure on the pistol to keep the rear sight secured, push downward on the pistol in one continuous motion to lock the slide to the rear.

  Note: Ensure the pistol does not move to safe when locking the slide to the rear.

- Place the pistol in a position to observe the chamber.
- Correct the stoppage.

If there is a round in the magazine but not in the chamber, release the slide and observe a round being chambered. If you do not observe a round being chambered, tap, rack, and bang to perform remedial action:

<u>Tap</u>

- Keep muzzle pointed in a safe direction and place trigger finger straight along the receiver.
- Strike the bottom of the magazine against a hard surface to ensure it is seated. In the kneeling position, strike the magazine against the thigh (see fig. 8-7 on page 8-6). In the standing position, bring the right knee up while striking magazine against the right thigh (see fig. 8-8 on page 8-6). In the prone position, strike the magazine against the deck.

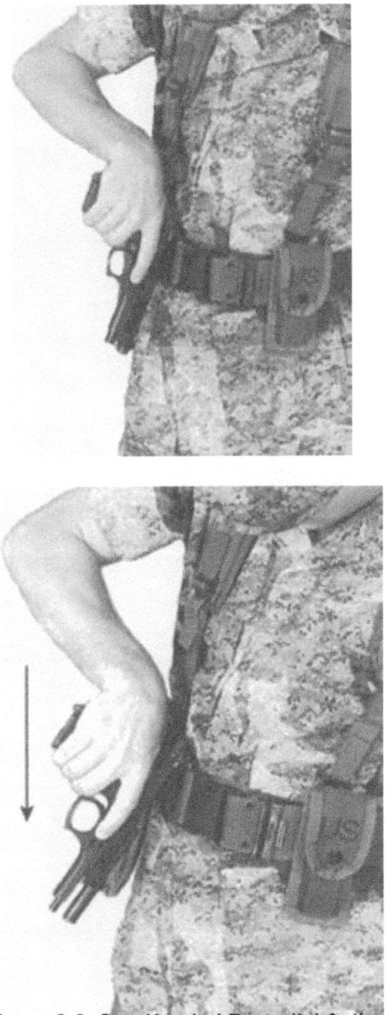

Figure 8-6. One-Handed Remedial Action.

Figure 8-7. Tap the Magazine Against the Knee: Kneeling Position.

Figure 8-8. Tap the Magazine Against the Knee: Standing Position.

Rack

- Rotate the pistol so the rear sight can be hooked on the top of the cartridge belt or any other surface (e.g., the edge of a table, wall, heel of the boot) that provides the resistance needed to rack the slide in a smooth, uninterrupted motion. Do not let the muzzle of the pistol cover the body. See figures 8-9 and 8-10.

Figure 8-9. Rack Slide Against Cartridge Belt.

Figure 8-10. Rack Slide Against Heel of Boot.

- Apply pressure on the pistol to keep the rear sight secured while pushing downward on the pistol to move the slide fully to the rear.
- Release pressure on the pistol to chamber the round.

Bang

- Recover the pistol on target, re-establish sight alignment/sight picture, and attempt to fire.

Presentation From the Holster With the Weak Hand

During combat, the Marine must be prepared to engage targets at any time. Therefore, the Marine must have the ability to present the pistol from the holster with his weak (left) hand if circumstances prevent him from using his strong (right) hand. He must apply the same smooth, controlled movement to present the pistol to the target. There are three methods for withdrawing the pistol from the holster with the weak hand. The Marine should practice each method to determine which works best for him.

Method One: Pistol Rotation

- Unfasten and release the D-ring with the left hand. See figure 8-11.

Figure 8-11. Method One: Release D-Ring.

- Wrap the fingers and thumb around the pistol grip so the index finger and thumb are around the base of the pistol grip. See figure 8-12.

Figure 8-12. Method One: Grasp the Pistol.

- Draw the pistol straight up until trigger guard clears the top of the holster. See figure 8-13.

Figure 8-13. Method One: Clearing Trigger Guard.

- Rotate the pistol so the magazine well faces the target and the trigger guard rests on the top of the holster. Ensure the trigger guard rests securely against the top of the holster. See figure 8-14 on page 8-8.

Figure 8-14. Method One: Rotating the Pistol.

- Maintain downward pressure on the pistol and slide the left hand around the pistol grip to establish a firing grip. See figure 8-15.

Figure 8-15. Method One: Establishing the Firing Grip.

- Grasp the pistol and continue to withdraw the pistol until the muzzle clears the holster and rotate the muzzle toward the target.
- Sweep the safety off with the thumb of the left hand while moving the pistol out toward the target. At the same time, place the trigger finger on the trigger and establish sight alignment and sight picture within the aiming area.
- Continue trigger pressure until the shot is fired.

Method Two: Hand Rotation

- Unfasten and release the D-ring with the left hand.
- Push the holster flap up and out of the way with the back of the left hand.
- Push the right hip out to facilitate grasping the pistol grip. Grasp the pistol grip between the holster flap and the pistol grip, the thumb is wrapped around the other side of the pistol grip against the backstrap. See figure 8-16.

Figure 8-16. Method Two: Grasping the Pistol.

Note: The cartridge belt may be pulled with the left hand to bring the holster closer to the body's center to facilitate grasping the pistol.

- Draw the pistol straight up while rotating the pistol so the magazine well faces inboard. Continue to remove the pistol until the trigger guard rests on the holster. See figure 8-17.
- Establish a firing grip and rotate the thumb to a position to operate the safety. See figure 8-18.
- Continue to withdraw the pistol until the muzzle clears the holster and rotate the muzzle to the target. See figure 8-19.
- Sweep the safety with the thumb of the left hand while moving the pistol out toward the

Figure 8-17. Method Two: Trigger Guard Rests on Holster.

Figure 8-18. Method Two: Establish Firing Grip.

target. At the same time, place the trigger finger on the trigger and establish sight alignment and sight picture within the aiming area.

- Continue trigger pressure until the shot is fired.

Method Three: Knee Placement

This method is particularly effective when firing from the two-knee kneeling position and the standing position.

Figure 8-19. Method Two: Rotating the Muzzle.

- Unfasten and release the D-ring with the left hand.
- Wrap the fingers and thumb around the pistol grip so the index finger and thumb are around the base of the pistol grip.
- Draw the pistol straight up until the muzzle clears the holster and rotate the muzzle forward. The magazine well should face upward.
- Place the pistol between the thighs, applying enough tension to keep the pistol in place. The Marine must ensure that he does not disengage the safety while performing this motion and that the muzzle of the pistol is clear of the body. In the standing position, the Marine bends slightly at the knees to facilitate securing the pistol between the thighs.
- Remove the left hand from the pistol grip and rotate the hand around the pistol grip to facilitate establishing a firing grip.
- Grasp the pistol grip and remove the pistol from the thighs, rotating the pistol until the muzzle is pointed toward the target.
- Sweep the safety with the thumb of the left hand while starting to punch the pistol out toward the target.
- Establish a firing grip and continue to punch the pistol out toward the target. At the same time, place the trigger finger on the trigger and

establish sight alignment and sight picture within the aiming area.

- Continue trigger pressure until the shot is fired.

Transferring the Pistol From One Hand to the Other

Situations may arise in combat where the Marine must transfer the pistol to the left hand to engage a target because the right hand or arm is injured. To safely transfer the pistol from one hand to the other, perform the following steps:

Note: This procedure is written to transfer the pistol from the right hand to the left hand.

- Point the muzzle in the direction of the target and place the trigger finger straight along the receiver. See figure 8-20.

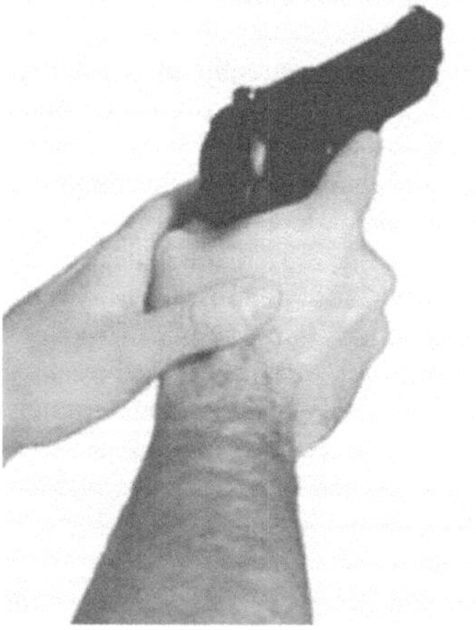

Figure 8-20. Transferring the Pistol—Step 1.

- Place the web of the left index finger and thumb under the backstrap of the pistol. See figure 8-21.
- Wrap the fingers of the left hand around the pistol grip while releasing the right hand grip.

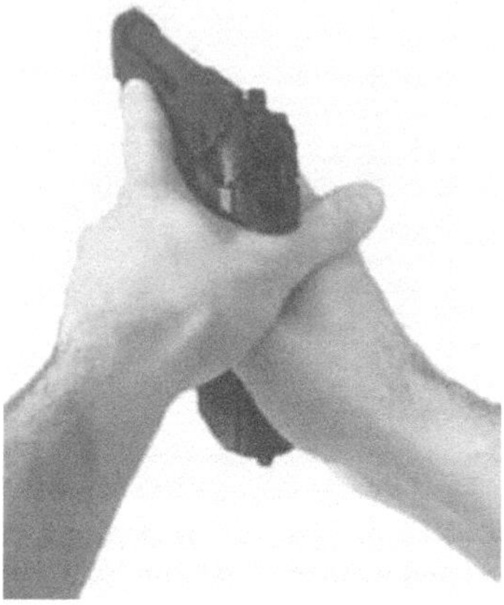

Figure 8-21. Transferring the Pistol—Step 2.

- Establish a firing grip with the left hand on the pistol. See figure 8-22.

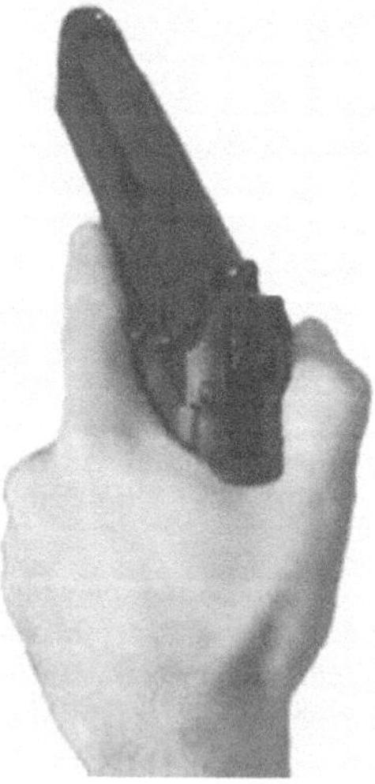

Figure 8-22. Transferring the Pistol—Step 3.

CHAPTER 9
ADVANCED TECHNIQUES

Some tactical situations are quite complex and require Marines to apply advanced pistol engagement techniques in order to effectively engage the enemy and survive. The ability of the Marine to rapidly shoot on the move and to turn and fire will be critical to a successful engagement.

> *Note:* The procedures in this chapter are written for right-handed Marines; left-handed Marines should reverse directions as needed.

Shooting on the Move

When moving from one area of cover to another, the Marine may engage a target that presents an immediate threat. In this situation, the Marine relies on accurate fire to serve as his cover.

Moving With the Pistol

The likelihood of encountering a threat dictates how the pistol is carried while the Marine is moving. But regardless of the method of carry, the pistol's muzzle is pointed in the same direction that the head and eyes are looking—the eyes and muzzle move as one.

When carrying the pistol in the Alert position, the pistol is tucked in close to the body, finger off the trigger. This position allows the Marine freedom of movement and still allows for a quick presentation of the pistol.

If there is a high probability of encountering a threat, the Marine carries the pistol in the Ready position while moving.

Moving: The Glide Technique

To shoot accurately while moving, the Marine cannot run or walk fast. A normal running movement produces too much bounce in the Marine's body and makes it extremely difficult to use the sights of the pistol or to achieve the needed level of stability for accurate shooting. Therefore, the proper movement technique is similar to a glide and allows for accurate target engagement:

- A lower center of gravity is achieved by keeping the knees bent and the upper body erect.
- The bent knees also serve to absorb the shock generated by movement.
- The feet are not lifted as high as in a normal walk or run, allowing obstructions to be kicked out of the way.
- The feet and knees point in the direction of travel. If there is a need to engage a target off to the side, the Marine rotates at the waist to move the upper body in the desired direction. The upper body shooting position does not change.

The glide is not a technique that can be used for a long period of time or for long distances. Rather, the Marine can move in a normal manner until it becomes necessary to engage a target, he then slows and assumes the glide.

Engaging Targets: Using the Pistol Sights

When the Marine fires the pistol, he must concentrate on the pistol's front sight. The adherence to the fundamentals of marksmanship becomes even more vital when both the Marine and the enemy may be moving.

Continuing to Move

A Marine continues to move no matter what happens. Continuing to move makes it difficult for the threat to engage the Marine and can be one of the Marine's best defensive assets. When moving to cover or to a different position, the Marine should not stop to engage a target.

Reloading and Stoppages

If a stoppage occurs or a reload is required while moving, the Marine's primary consideration is to keep moving and seek cover. Ideally, reloads and clearing stoppages are performed behind cover.

Turn and Fire

The techniques for turn and fire allow a Marine to engage a target that is not directly in front of him. The key to turn and fire is smoothness and quickness of pivoting and presentation of the pistol to engage the threat. Turn and fire is based on the principle that wherever the head goes, the body follows.

Engaging Targets 90 Degrees to the Right and Left

The Marine turns the head toward the threat and identifies the target. Once the target is identified, the Marine maintains focus on the target for the rest of the presentation.

If engaging a target 90 degrees to the right, the Marine turns toward the target by raising the left foot while pivoting on the ball of the right foot. The Marine plants the left foot once he is facing the target squarely. The stance is natural. See figure 9-1. The Marine then engages the target.

Figure 9-1. Engaging Targets 90 Degrees to the Right.

If engaging a target 90 degrees to the left, the Marine turns toward the target by raising the right foot while pivoting on the ball of the left foot. The Marine plants the right foot once he is facing the target squarely. The stance is natural. See figure 9-2. The Marine then engages the target.

Engaging Targets 180 Degrees to the Rear

The Marine turns the head toward the target, looking over either the right or left shoulder, and identifies the target. Once the target is identified, the Marine maintains focus on the target for the rest of the presentation. See figures 9-3 on page 9-4.

If the Marine pivots to the right to engage the target, pick up the left foot, move it across the right foot, while pivoting 180 degrees on the ball of the right foot, plant the left foot, and squarely face the target in a natural stance. See figure 9-4 on page 9-4.

If the Marine pivots to the left to engage the target, pick up the right foot, move it across the left foot, while pivoting 180 degrees on the ball of the left foot, plant the left foot, and squarely face the target in a natural shooting stance. See figure 9-5 on page 9-4.

The Marine then engages the target.

Figure 9-2. Engaging Targets 90 Degrees to the Left.

Right Shoulder

Left Shoulder

Figure 9-3. Engaging
Targets 180 Degrees to the Rear.

Figure 9-4. Engaging Targets
180 Degrees to the Rear (Right Shoulder).

Figure 9-5. Engaging Targets
180 Degrees to the Rear (Left Shoulder).

CHAPTER 10
LOW LIGHT AND DARKNESS TECHNIQUES

An effective combat marksman must be prepared to detect and engage targets under a variety of conditions. Factors such as terrain and opportunity often dictate that the Marine engage combat targets at night or under low-light conditions.

The fundamentals of marksmanship are employed for engagement of targets in low light and darkness just as they are in daylight. However, the principles of night vision and target detection when engaging targets at night must be applied.

Combat Mindset

In the stress of the combat environment, Marines must eliminate any hesitation, fear, or uncertainty of action and focus on the actions to fire well-aimed shots. This is important during low light and darkness when attention is more easily diverted because the sense of vision is reduced. This may create a shock of awareness as the Marine relies more heavily on his other senses. Every noise, movement, and muzzle blast appears intensified at night. The physical acts of shooting must become second nature to the Marine, so his focus will not be diverted from firing well-aimed shots. This is accomplished through mental preparedness and training until shooting actions become instinctive.

Night Vision Adaptation and Maintenance

Night Vision Adaptation

There are two methods for acquiring night vision.

The first method is to remain in an area of darkness for about 30 minutes. This area can be indoors or outdoors. The major disadvantage of this approach is that a Marine is not able to perform any tasks while acquiring night vision in total darkness.

The second method is to remain in a darkened area under a low intensity red light (similar to the light in a photographer's darkroom) for about 20 minutes, followed by about 10 minutes in darkness without the red light. This method produces almost complete night vision adaptation while permitting the performance of some tasks during the adjustment period.

Night Vision Maintenance

Because the eyes take a long time to adjust to darkness, it is important to protect night vision once it is acquired. To maintain night vision, the Marine avoids looking at any bright light and shields the eyes from parachute flares, spotlights, or headlights. When using a flashlight to read a map or other material—

- Put one hand over the glass to limit the area illuminated and the intensity of the light.
- Keep one eye shut to reduce the amount of night vision lost.
- Cover the light with a red filter to help reduce the loss of night vision.

Target Detection Techniques

Once the Marine has acquired night vision, he is prepared to locate targets. Some of the daylight observation techniques, such as searching for target indicators, also apply at night or in low light. But night observation techniques must allow for the limitations of night vision and the need to protect it.

Off-Center Vision

Because of the placement of the cones in the center of the retina and the rods around the edges, the angle at which the Marine observes an object at night affects how well it is seen. Off-center vision is the technique of keeping the attention focused on an object without looking directly at it. To maximize the use of the eye's rods that provide night vision—

- Never look directly at the object being observed.
- Look slightly to the left, right, above, or below the object.

Each Marine must experiment and practice this technique to find the best off-center angle. Typically, the best off-center angle is 6 to 10 degrees away from the object or about a fist's width at arm's length. See figure 10-1.

Scanning/Figure Eight Scan

Scanning is the use of off-center vision to observe an area or object and involves slowly moving the eyes in a series of separate movements across the objective area. See figure 10-2. To execute a figure eight scan, move the eyes in a figure eight pattern in short, abrupt, irregular movements over and around the area. Once a target indicator has been detected, focus should be concentrated in that area, but not directly at it. It is more effective to scan from a prone position or a position closer to the ground than the object being observed, creating a silhouetted view of the object. When scanning an area, look and listen for the same target indicators as in daylight: movement, sound (which seems louder at night than during daylight), and improper camouflage. Also, objects in bright moonlight/starlight cast shadows just as in sunlight.

Effects of Illumination

Both ambient (natural) light and artificial illumination can affect a Marine's perception of the target's distance and size and his night vision. The following situations produce less of an effect at the close ranges that pistol engagements occur, but they still affect target detection and engagement:

- Light behind or between the Marine and the target illuminates the front of the target and makes it appear closer than it actually is.
- Light beyond the target displays the target in silhouette and makes it appear farther away than it actually is.
- The introduction of light requires the eyes to make a sudden, drastic adjustment to the amount of light received, causing a temporary blinding effect that occurs when night vision is interrupted abruptly. Ambient light can also have the

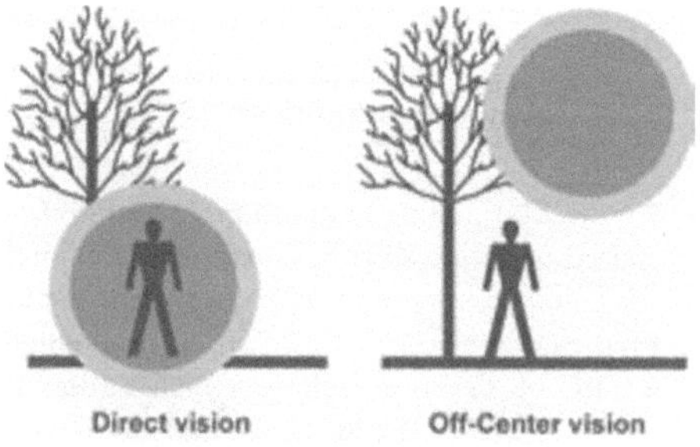

Figure 10-1. Off-Center Vision.

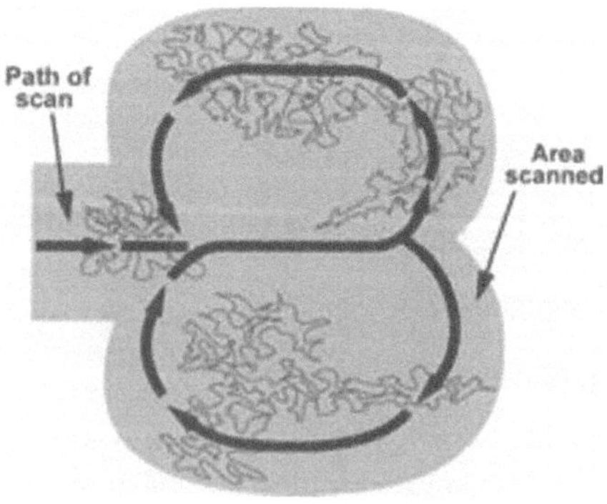

Figure 10-2. Figure Eight Scan.

same blinding effect; e.g., when a bright moon suddenly appears from behind clouds.

Acquiring Targets at Night

The Marine must keep both eyes open to get maximum visual coverage of the target area. Keeping both eyes open also improves depth perception and peripheral vision.

Sight Alignment/Sight Picture

Sight alignment and sight picture are obtained the same way in darkness as in daylight. There is normally enough ambient light to perceive objects as far away as 50 meters, especially if they are moving. However, when pistol sights are placed on a dark background, such as a camouflaged target, the Marine may not be able to acquire and align the sights clearly. Instead, the Marine may have to rely almost entirely on his presentation, firing position, and grip to get the pistol on target. To check for sight alignment or acquire the sights:

- Move the pistol just off the edge of the target to an area that provides a good contrast.
- Acquire sight alignment.
- Bring the sights back on line with the target while applying pressure to the trigger so the shot breaks once the sights are on target.

Flashlights

A flashlight can be used in low light and darkness to identify and illuminate targets. It is also used to acquire sight picture.

Types of Flashlights

There are a variety of flashlights used throughout the Marine Corps. They come in many different shapes and sizes; however, they are either straight/tubular or L-shaped.

Straight/Tubular

The straight/tubular flashlight is the most common. The body of the flashlight is a straight/tubular shape with the lens at the head of the flashlight. The on/off mechanism is located along the body or at the base of the flashlight.

L-Shaped

The L-shaped flashlight has a straight/tubular body, with the head/lens of the flashlight positioned at a 90-degree angle to the body of the flashlight. The on/off mechanism is located along the body or at the base of the flashlight.

Activation Devices

Depending on the type of flashlight, there are several ways that it can be activated.

On/Off Button

On/off buttons are located either along the body of the flashlight or at the base. On/off buttons are operated in a number of ways, depending on the flashlight.

On/Off Switch

On/off switches are mounted on the body of the flashlight. This switch has three settings that allow the flashlight to be either on, off, or in manual mode. In the manual mode, the flashlight is turned on by pressing and holding the button located next to the on/off switch.

Rotating Head

Rotating the head of the flashlight, either clockwise or counterclockwise, turns the flashlight on. Continuing this motion allows the width and intensity of the beam to be adjusted.

Types of Lenses

Clear Lenses

Clear lenses are used primarily for target detection and to illuminate the pistol sights. They are the most commonly used lens.

Colored Lenses

Colored lenses include red, amber, and blue. These lenses are used primarily for map reading and signaling, but can be used to illuminate the pistol's sights.

Target Detection

When a target is illuminated, the front sight may become silhouetted against the target, providing the sight picture needed to engage the target. During target detection, the Marine's focus should be twofold: to scan the area to identify possible targets and to assess the area to formulate a plan for engagement or cover (e.g., identifying the quickest route to cover, determining the best method for engagement based on terrain).

Searches

The Marine executes a hasty search immediately upon entering a new area in order to identify a threat that poses immediate danger. To execute a hasty search—

- Quickly scan the area with the flashlight taking note of obvious points throughout the area that could cover or conceal the enemy. Wherever the eyes move, the flashlight should move.
- Aim the flashlight beam on the ground about 8 to 10 feet in front of the Marine's location, this allows the eyes to follow the beam and quickly establishes a reference point for the light. Aiming the beam at the outer edge of the search area strains the eyes to find the beam and decreases the field of view.

The Marine executes a detailed search after the hasty search. A detailed search focuses on target indicators identified during the hasty search. To conduct the detailed search—

- Focus the most direct or intense portion of the flashlight beam on the target indicators.
- Move the beam of light slowly across the target indicator from right to left or left to right. Wherever the eyes move, the flashlight should move.

Considerations

During night operations, the Marine should use a flashlight whenever possible. A flashlight not only helps the Marine locate targets at night, but the beam can be aimed directly at the target once it is detected and temporarily blinds the target. This gives the Marine the advantage to react before the target does. However, the Marine must consider the following:

- Anytime a flashlight is on, the location may be revealed to the enemy. The Marine should keep the flashlight pointed out in front of his body to help avoid illuminating himself and revealing his position.
- Light shined directly from in front of the Marine at the target obscures the Marine.
- Light shined from the side of the Marine at the target illuminates the Marine.
- Light can bounce off surfaces and reflect back onto larger areas that the light is not focused on. Therefore, whenever a Marine shines a flashlight onto a surface, some of the light may reflect back and illuminate the Marine.
- When positioned at the outside corner of a room, building, or other cover, the Marine should avoid pointing the flashlight beam directly at the corner because the beam reflects off the corner and illuminates the Marine. To prevent this, the head of the flashlight should extend just beyond the corner.
- Many flashlights have adjustable beams that aid in target detection. The beam's intensity must be adjusted to provide the best illumination of the area depending on the distance between the Marine and the area of observation.
- Diffused light from a wide beam creates a softer light and illuminates a greater area, but the beam does not travel a great distance. A wide beam is best for observing larger areas at close range.
- Concentrated light from a narrow beam illuminates a smaller field of view, but the beam travels a greater distance. A concentrated beam is effective for observing a specific area or an area that is further away. Concentrated light

blinds the enemy and prevents him from focusing on the Marine or determining his location.

- When searching an area, the Marine can alter the position where he is holding his flashlight so that the beam is perceived as coming from various locations. For example, the Marine can turn the flashlight on from a standing position, quickly search for targets, turn the flashlight off, assume a kneeling position, and search again. This keeps the enemy from obtaining an exact location of the Marine's position.
- Keep both eyes open to get maximum visual coverage of the target area. Keeping both eyes open also improves depth perception and peripheral vision.

Target Engagement

Once a target is detected using a flashlight, the flashlight can be used to acquire sight picture and facilitate engagement of the target. The fundamentals of marksmanship are employed for engagement of targets in darkness just as they are in daylight. However, the Marine must apply the principles of night vision and target detection and must be able to employ a flashlight properly when engaging targets at night.

Grip

To engage a target accurately while using a flashlight, the flashlight must be held to maintain the pistol's stability, control, and recoil during firing. Normally, the left hand (two-handed grip) provides the pistol's stability and ability to manage recoil. However, when firing with a flashlight, the left hand holds and operates the flashlight. Therefore, some stability, control, and management of recoil is lost.

Securing the Flashlight

Most flashlights come equipped with a retaining loop located at the base for ease of transportation. To position the cord properly:

- Feed one end of the cord through the retaining loop of the flashlight and tie the ends of the cord together.
- Slip the left hand through the cord loop. The loop should be just big enough so that, if the flashlight needs to be dropped, it can be retained on the wrist.
- Rotate the flashlight until the cord is twisted to the desired tension and length necessary to best stabilize the rear end of the flashlight in the hand.
- Grasp the flashlight with the left hand in a position that allows it to be operated easily.

If the flashlight has a retaining loop, attaching a cord to the loop helps support and stabilize the flashlight when the Marine must fire the pistol with the flashlight in his left hand. The cord also helps retain the flashlight.

Acquiring Sight Alignment/Sight Picture

When holding a flashlight on a target, the tendency is to look at the target rather than the sights. Sight alignment is still necessary for effective target engagement. Using a flashlight to illuminate a target allows the Marine to—

- Acquire sight picture by silhouetting the sights against the target.
  - The Marine establishes sight picture by focusing the sights in the soft, diffused light area of the target.
  - At close ranges, colored lenses produce enough light on the target to silhouette the sights; at long ranges, colored lenses will not provide enough light off the target to illuminate the pistol sights. The Marine must be able to see the target by some other means, (e.g., contrast, moonlight).
- Acquire sight picture by illuminating the sights.
  - A colored lens can be held directly over the sights to shine light on the top of the pistol to illuminate the sights. This allows sight alignment to be established without revealing the Marine to the enemy. This method can also be used with a clear lens to make a precision shot at long ranges but the Marine risks being illuminated by the light.
  - A colored lens can be shined from directly behind the pistol sights to illuminate them. This

allows sight alignment to be established, but reveals the Marine's position to the enemy.

Target Engagement Techniques

Two-Handed Grip Technique. This technique is typically used with a straight/tubular flashlight. The pistol and flashlight must be side by side and level so the Marine can engage a target without making adjustments to the pistol or the flashlight. This technique is also effective for firing multiple shots because the flashlight and pistol recoil as a unit.

Whenever possible, the flashlight is just in front of the muzzle of the pistol so that it does not illuminate the pistol. Placement of the flashlight alongside the pistol may need to be adjusted depending on the size of the flashlight or the location of the on/off mechanism.

This technique is most often used with an Isosceles position (see fig. 10-3):

- Grasp the flashlight with the left hand, lens pointing down range.
- Wrap the thumb and index finger of the left hand around the body of the flashlight, thumb resting on the on/off switch to allow easy operation of the on/off switch without disrupting the grip on the flashlight.
- Extend both arms toward the target and bring the flashlight alongside the pistol so the fingers of the left and right hands touch.
- Wrap the bottom three fingers of the left hand around the fingers of the right hand, incorporating the flashlight into a two-handed grip.
- Apply isometric pressure against both hands to aid in stabilizing the pistol.

This technique can also be incorporated into a Weaver position; the more the left hand wraps around the right, the more the body can be angled and the left elbow bent. Apply push-pull pressure on the grip to stabilize the pistol. See figure 10-4.

Cross-Hand Technique. The cross-hand technique is used primarily with a Weaver position and can be used with either a straight/tubular flashlight or an L-shaped flashlight (see fig. 10-5):

- Grasp the flashlight with the left hand, fingers wrapped around the top of the flashlight and thumb wrapped around the bottom.

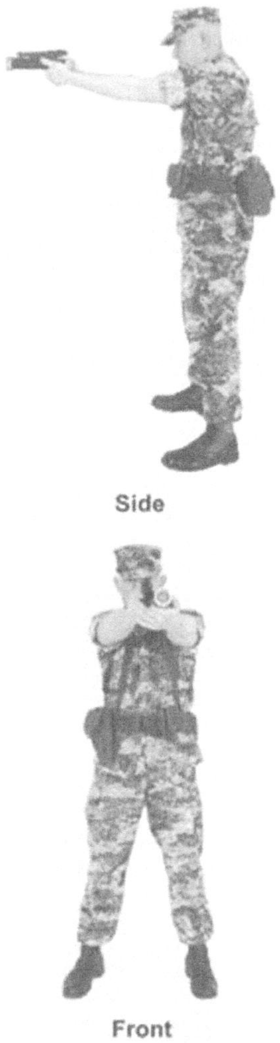

Figure 10-3. Two-Handed Grip Technique With Flashlight (Side and Front View).

Figure 10-4. Two-Handed Grip Technique With Flashlight (Weaver Position).

Figure 10-5. Flashlight Cross-Hand Technique.

- Place the thumb on the on/off button while maintaining a firm grip on the flashlight.
- Extend both arms toward the target and bring the left hand under the pistol so that the back of the left hand is resting firmly against the back of the right hand.
- Apply isometric pressure against both hands to aid in stabilizing the pistol.
- Extended the right arm fully and bend the left arm at the elbow.

Considerations for the Carry/Transport

When searching an area for targets, the Marine moves with the pistol at the carry or transport dictated by the threat level.

The Ready. If enemy contact is expected (contact imminent), the Marine carries the pistol at the Ready and carries the flashlight in his left hand and incorporates it into his firing grip. In the Ready, the Marine can search for targets and readily present the pistol and flashlight to the target for engagement without making adjustments to either the grip or the flashlight. See figure 10-6 on page 10-8.

The Alert. If enemy contact is likely, the Marine carries the pistol at the Alert and the flashlight in his left hand and incorporates it into his firing grip. The Marine must ensure the flashlight is not pointed at the deck because it illuminates the Marine. If necessary, the head of the flashlight may have to be tilted up to elevate the beam to increase the field of view. If the flashlight has been tilted, it has to be lowered so it is level with the pistol muzzle and the light shines directly on the target in order to establish sight picture. See figure 10-7 on page 10-8.

Holster Transport. If there is no immediate threat, the pistol is holstered and the Marine carries the flashlight with his left hand to search for targets. Should a target present itself, the Marine presents the pistol from the holster and engages the target. Time and distance to the target dictate whether the Marine incorporates the flashlight and his left hand into the firing grip.

Cross-Hand Two-Handed Grip

Figure 10-6. Flashlight Ready Carry (Cross-Hand and Two-Hand Grip).

Two-Handed Grip Cross-Hand

Figure 10-7. Flashlight Alert Carry (Two-Handed Grip and Cross-Hand).

Made in United States
Cleveland, OH
05 November 2026

47673671R00123